SCC

D1241649

The Two Principal Laws of Thermodynamics

The Two Principal Laws of Thermodynamics

A Cultural and Historical Exploration

J. H. van den Berg

Introduction by Bernd Jager

translated by Bernd Jager, David Jager & Dreyer Kruger

Duquesne University Press
Pittsburgh, Pennsylvania

First published in Dutch as *Twee wetten: De twee hoofdwetten van de thermodynamica*, © 1999, Uitgeverij Pelckmans, Kapelsestraat 222, 2950 Kapellen.

Published in the United States of America by:
Duquesne University Press
600 Forbes Avenue
Pittsburgh, Pennsylvania 15282

Library of Congress Cataloging-in-Publication Data

Berg, J. H. van den (Jan Hendrik), 1914–
[Twee wetten. English]
The two principal laws of thermodynamics : a cultural and historical exploration / J. H. van den Berg ; translated by Bernd Jager, David H. Jager, and Dreyer Kruger.
p. cm.
Includes bibliographical references (p.) and index.
ISBN 0-8207-0354-0 (cloth : alk. paper) — ISBN 0-8207-0355-9 (pbk. : alk. paper)
1. Thermodynamics. I. Title.
QC311.B4813 2004
536'.7—dc22

2004001628

∞Printed on acid-free paper

Dedicated to the memory of
Professor Ed Murray, C.S.Sp.,
"Apostle of Imagination"

Contents

The Historical Background of van den Berg's *Two Laws*

About the Author

It is a rare occasion to read an essay on a historical development in physics written by a historian who is also a noted authority on psychiatry and psychology and who, moreover, has made it his lifework to develop an understanding of the human condition inspired by the work of Gaston Bachelard, Husserl and Heidegger.

Van den Berg is the author of some 35 books on topics ranging from architecture to entomology and from psychotherapy to theology. Several of these have been translated into English and into half a dozen other languages and many have seen numerous reprints. His books *The Changing Nature of Man* (1961) and *Medical Power and Medical Ethics* (1978) became international best sellers. All are written in a very direct and vivid style that confronts the reader with the sharp contours of a human situation or the outlines of a particular problem but which also opens a vast terrain for new thought. The ostensible clarity of the texts and the attractive illustrations serve to attract, if not seduce the reader. But once caught up in the text the reader is confronted by subtle enigmas and introduced to labyrinths from which it is impossible to escape unchanged in spirit and in mind.

A recurrent trait of van den Berg's historical writing is his capacity to uncover relationships between the most heterogeneous cultural elements of a particular period so that they begin to form a comprehensive and mutually clarifying whole. His view of a historical period is symphonic. A dominant theme plays itself out in architecture and is then taken up and elaborated by a political theory, after which it may gain new life in a particular fashion design or in a scientific discovery or a breakthrough in mathematics. All these varied cultural activities appear in this work as so many instruments contributing to the creation of a symphonic whole. To discover this whole we are required to assume the role of a good audience that has come to observe, to listen and to bear witness to a bygone age. The coherence and the unity we find or fail to find in the world of history is ultimately a function of our ability to properly address an era and to make it speak to us and reveal itself.

The author's long and productive academic career spans more than half a century. He was born on the eve of the First World War and grew up amid the political and ideological ferment of the years preceding the Second World War. He began his professional life as a teacher of mathematics, but then decided to study medicine and to specialize in psychiatry. He became chief assistant to the famous Dutch psychiatrist H. C. Rümke at the University of Utrecht psychiatric clinic. Having completed his formation as a psychiatrist he traveled to Paris, where he participated in the postwar philosophical and literary renaissance that was then in full swing. He became part of the circle of Henri Ey and Jean Wahl and studied with Gaston Bachelard, who influenced his later work. In turn, Bachelard was influenced by his younger friend. He cited van den Berg's *The Phenomenological Approach in Psychology* in his introduction to *The Poetics of Space* and referred to the author as "this learned Dutch phenomenologist." Subsequent to his studies in France, van den Berg worked in the clinic of Manfred Bleuler in Switzerland and met Ludwig Binswanger. When he returned to Holland after his extended study tour, he opened a private practice in Utrecht and shortly thereafter began his academic career as a lecturer in psychopathology at the University of Utrecht. In 1951 he obtained a chair in pastoral psychology and three years later joined the University of Leiden and taught phenomenological psychology and psychiatry.

Most of van den Berg's early writing dealt with medical and psychiatric topics, but his phenomenological orientation pulled him inexorably into the orbit of an interdisciplinary understanding of human situations and in the direction of a philosophical anthropology. His inaugural address on the relationship between psychology and religion opened a lifelong dialogue with theology that led to the publication in 1995 of his *Metabletica van God* (Metabletics of God). *The Changing Nature of Man* was published in Dutch in 1956 and in English in 1961. It makes a significant contribution to psychology and psychotherapy, but it can also be read as a treatise on education, history and sociology. The text transcends the limits of any particular disciplinary context and can best be appreciated as a series of great poetic meditations on the human condition that grows more enchanting and revealing with each subsequent reading.

Yet another interdisciplinary study, *Het Menselijk Lichaam* (The human body), was published in the late fifties. The book explores the historical transition from ancient to modern medicine and from a traditional exploration of the "closed" human body to a modern anatomical and physiological exploration of the "opened" body.

To do justice to this intellectual revolution within medicine, van den Berg felt compelled to move beyond the limits of medicine to explore the larger historical world in which this revolution took place. This larger world was being transformed by the discovery of the New World at the same time that it was being assailed by new religious sensibilities, by changed relationships between the sexes, by different ways of dressing and behaving and by new trends in the fields of literature, painting and architecture. In reading these texts one learns that it is not possible to gain a proper insight into the historical developments within a particular discipline without placing these within the context of a larger cultural world. To understand a particular breakthrough in medicine or physics we must bring it into a relationship to broader historical currents that affect human relationships, to self and other, as well as to the material world.

Van den Berg's phenomenological history or "metabletics" refuses us the protection of a particular academic niche but forces humanists and human scientists to interact within a wider cultural landscape. It does not permit us to enclose ourselves within the safe cocoon of our own time and circumstances in the mistaken belief that our own contemporary convictions and hobbyhorses can safely take the place of all that was thought and understood before our time. It disabuses us of the phantasms of a progressive history that automatically registers improvements as it steadily moves from an inferior past toward its crowning achievement in the present. Quite to the contrary, van den Berg's metabletic history invites risky encounters with distant times and places and with sensibilities and achievements that are neither inferior to, nor merely anticipatory of our own.

Metabletics, Metabolism and Metaphor

Van den Berg's essay on the two laws of thermodynamics makes use of what the author has variously described as a phenomenological or "metabletic" approach to the writing of history. The term *metabletics* is coined from the Greek verb *metaballein*, which means "to change," "to turn around" and "to alter." It refers to an abrupt change in one's goals, one's direction or one's way of life. It implies an abrupt change in which the new state of affairs leaves little or no place for the old state of affairs and in which the present absorbs or radically distances itself from the past. Our word *metabolism* still evokes this abrupt and complete change undergone by plants and animals as they are eaten and digested and as they are made to form part of the body of an alien organism. In this process of metabolism the "old" life form loses every trace of its former identity as it is gradually being transformed into a "new" life form. *Metaballein* refers literally to the act of throwing (*ballein*) something across a spatial or temporal interval (*meta*), in such a way that the path of transformation is lost and cannot be traversed in an opposite direction. Metabolism refers to a fateful, unidirectional change that leaves no memory of its passing.

In that sense it evokes most clearly the image of death, understood here as an absolute and irrevocable loss of one's identity and of one's place in the world. To better understand this fateful metabolic transformation we need to contrast it with an opposite trend that preserves identity and places it on a new foundation.

We might compare the Greek verbs *metaballein* and *metapherein* and their English derivatives, *metabolism* and *metaphor*. Both *metapherein* and *metaballein* refer to a displacement or transport from one site to another. Both invoke the passage of time and imply a "before" and an "after," in radically different ways. Both imply the transport of something from one context or one world to another.

Meta-pherein and *meta-phor* refer to a transport across a threshold or a bridge that interconnects two separate and distinct, but never isolated or mutually exclusive, realms. The Greeks used the verb to refer to money management, to transferring funds from one account or one particular purpose to another. In horse racing it referred to the alternative use of the goad, first to one and then to the other of a team of horses. Within the context of bureaucracy it referred to the transfer of officials from one post to another, while in rhetoric it could point to the displacement of a particular word from a known to an unknown or novel context. In the art of translation it referred alternatively to the Greek word or concept and to its equivalent in the foreign tongue.

Meta-pherein means literally "to carry something from one place to another." *Meta-pherein* and *meta-phor* create a path that links two domains via a threshold so that it can be traveled and retraveled in both directions. The word that has been translated in another language can be retranslated in the original tongue; the official assigned a different post can thereafter be reassigned to his previous one, and funds allocated for a different purpose can subsequently be reassigned to their previous uses.

Meta-ballein and *meta-bolism* refer to a very different type of movement that does not honor the thresholds that protect the identity of neighboring domains. Instead, it destroys all distance and difference between them as it turns the one into the other. This dissolving movement of metabolism poses a constant threat to the human world that can be met only with the help of metaphor. It is for this reason that the movement implicit in metabolism signals the end of history unless it is contextualized and contained by the work of metaphor. Metaphor opens history since it creates a threshold and a path leading back and forth from one temporal or spatial domain or from one historical period to a preceding or succeeding one.

To read a story means to participate in a metaphoric transport that links the beginning to the end and the end to the beginning. It is this constant mutual reference of beginning and end that preserves the identity and the integrity of the narrative characters and events as they undergo changes in the course of the story. Narrative fails when it does not link a beginning to an end or when it does not succeed in preserving the identity of the characters throughout the course of the story. Narrative also fails when the reader does not observe the courtesies of the threshold and neglects to maintain a proper metaphoric distance from the characters and situations in the narrative. Reading becomes distorted either by identifying too closely with the characters, or alternatively, by failing to enter into a meaningful relationship with them.

Metaphor traverses a distance; it moves something, a character, a thing, a situation across a threshold or from one to the other side of a river in such a way that the path it follows forms a bridge linking two shores. Metaphor navigates between two different worlds and unites them into a meaningful whole. Metabolism, on the other hand, erases the distance between two different worlds. It collapses narrative and ends history as it effaces the difference between a beginning and an end. Metabolism is a ship that makes the river disappear as it crosses from one shore to the other; it burns the bridges built by metaphor. Metaphor gives and metabolism bars access to an intersubjective world of neighborliness, of conversation and storytelling. Van den Berg's history begins with the discovery of a metabletic, sudden and ineluctable shift and it then seeks to restore that change to the realm of metaphor.

We think here of metaphor as the governing principle of the cosmos and the lived world since it establishes a threshold between two different worlds. It creates a time and a space within which friendships can be formed, love can blossom, families can become established and cities can be built. Metaphor creates a world in which a host can meet a guest, a stranger can meet a native, a man can encounter a woman, and a child can come to love and trust an adult. It creates a world where the living may remember the dead, where mortals can address immortals and where heaven and earth do not collapse into one another but form together a meaningful whole. Metabolism, when unopposed and unrestrained, moves in the opposite direction; it creates a literal and material unity that recognizes nothing beyond itself and accepts no law or rule beyond those of natural necessity. As such it opens upon a world of pure violence in which there is no room for thresholds or for moral law.

We think of the lived world as a world of metaphoric couples, that is, as the dwelling place of neighbors, hosts and guests, natives and foreigners, the living and the dead and gods and mortals. By contrast, we think of the modern natural scientific universe as forming a severe, material and literal unity. In the place where the lived world shows us the historicizing and metaphoric unity of couples the universe shows us an ahistoric and metabolic unity of mutually devouring objects and forces.

The scientific study of a natural universe sheds a unique and revealing light on the lived world we inhabit. Yet we should not confuse the two or reduce the one to the other. To approach the universe as origin and destiny of our lived world or to approach natural science as the ultimate harbor of our thought can only undermine and ultimately destroy the world in which we actually live our lives. It means to accept metabolic change as the ultimate truth of our world and to ignore that the unity of a human world is metaphoric and applies to couples.

Both metabolic and metaphoric change form part of the human world. Death, disease, suffering, stupidity and forgetfulness all remind us of our fragile and mutable existence. They all reveal that the human body is destined to disappear without remainder into the body of the earth. Yet every threshold that links separate domains, that safeguards and treasures the distinct qualities of those who cross it, reminds us of a very different, specifically human way of joining and bringing together. Every mistake we make, every stumble on our path reminds us of the metabolic truth of a biological death. But every act of love and friendship, every thoughtful remembrance of the dead, every prayer or sacrifice offered to the gods and every attempt to transmit our cultural treasures from one generation to the next reminds us of a different truth. It is this latter truth that holds our world together. Every sound work of art, every well-made and useful thing, every story told or written and every work of science or technology that illuminates our world points beyond the merely metabolic truth of death and decay in the direction of a world made lucid by metaphor. Humanity cannot maintain itself without extending a welcome to the past and to the future. It can prosper only when it cultivates thresholds that bind together an inside and an outside, and that can both keep distinct and hold together a past, a present and a future. Both metabolic and metaphoric ways of joining play an important role in our life. Yet we all recognize instinctively that metabolism needs to be contained by metaphor. It is for that reason that our social customs embed the act of eating and digestion within a larger metaphoric context of hospitable encounters and table conversation. To eat or drink outside that context incurs the danger of becoming swallowed up in a metabolic world of compulsion and perversion. Our history contains periods of radical or metabolic change where for a moment all vital contact seems lost between a present and a preceding way of life. Van den Berg's metabletic history searches out such moments of radical historical change when a present time appears to have devoured its own past and obliterated the traces of a previous way of life. What is lost at these moments is the very possibility of a dialogue with an alternative way of understanding our world. We may think of metabletic history as an effort to restore to a hermetic present the metaphoric unity of past, present and future.

In the essay that follows, van den Berg describes such a radical change that took place towards the end of the eighteenth century. This revolutionary change not only sought to erase differences between individuals, classes and functions, it also eroded distinctions between animals and human beings and between living beings and inanimate things. These radical changes severed relations with preceding ways of approaching and understanding the natural world. It is in this changed climate of revolutionary thought that it became possible to formulate the laws of thermodynamics.

In the Beginning Was the Word

What is noteworthy about van den Berg's historical investigations are the range of the historical facts and situations he manages to assemble into a meaningful narrative whole. As modern readers we are accustomed to works that follow historical developments within a particular cultural domain such as art, architecture, economics or political ideologies. Only rarely do we come across a work that makes an attempt to link advances in chemistry to political ideology or that brings trends in ladies' fashions into a meaningful relationship to an ongoing military conflict.

We are accustomed to thinking about our cultural life as intrinsically changeable and so we are not surprised to learn that Renaissance architecture differs in essential aspects from Gothic or Baroque architecture, and that Elizabethan tragedy follows principles that are not applicable to classic Greek or French tragedy. We also readily accept the fact that the natural sciences themselves are subject to historical change so that a sixteenth century conception of material reality is not directly comparable to that of the seventeenth or the twentieth century. Yet we think about the natural and physical world itself as governed by immutable natural laws and as being beyond the reach of historical change.

Van den Berg enlarges our notion of what can become subject to historical change, expanding it to include material substances such as water and iron, wood and bread. He refuses to relegate the material world to some ideal domain wholly beyond the shaping power and creative ferment of an ongoing human and divine conversation. He introduces us to a historical world in which the material substances themselves and not merely their appearances are subject to historical change. Historical change refers here not merely to a temporal and material rearrangement of the world but to an ongoing, constantly shifting human and divine conversation that binds a mortal human world to a world beyond. The author refers to the elemental change in the material world that took place toward the end of the eighteenth century as "transubstantiation."

The revolutionary struggle for social and political equality in the eighteenth century did not remain limited to the political or social sphere but had its repercussions within the material world as well. Van den Berg points out that prior to the French Revolution social inequality had not been perceived as a malleable social fact but as an inalterable feature of the human landscape. This acceptance of inequality was not limited to the social realm or to interpersonal relations; it also formed part of the natural and material landscape. The eighteenth century looked upon water, bread or wood as local products that manifested the particular characteristics and qualities of the landscapes and localities of which they formed a part. The water of the Thames, the Danube or the Seine were therefore not directly comparable to one another, in the same way that the marble from Mt. Pentelicos in Attica was not quite the same as that from Paros in the Cyclades. Nor was it always a self-evident truth that the bodies of human beings born and raised in one part of the world could be directly compared to the bodies of people from other places and regions.

The story is told that when the famous British physician Thomas Harvey presented his discoveries about the function of the heart and the circulation of blood to a distinguished audience of German physicians he was applauded for his efforts but also firmly told that what held true for the bodies of Englishmen did not apply to the bodies of Germans. The idea of a universal medicine based on a generic human body at one time sounded as strange and improbable as the proposition that it was the selfsame and identical chemical compound water that coursed through all the rivers of the world.

Van den Berg reminds us that Lavoisier's *Traité élementaire de chimie*, the book that introduced the idea of universal chemical compounds, was published in the year 1786 on the eve of the French Revolution and at the time of the storming of the Bastille. It made its appearance at the very moment when the ideas of a classless society and of a universal humanity began to take hold of the body politic. The author points out that the idea of "equality" formed the core of the revolutionary slogan "Liberty, equality and brotherhood" and stood at the center of a new political doctrine and a novel anthropology. This idea extended its influence far beyond the social and political realm and also influenced our understanding of the physical and material world. Or perhaps it was the other way around: a new appreciation of universal and material relations began to extend its influence within the social realm.

It is evident that a historian whose work is guided by principles derived from the natural sciences will write a different kind of history than a colleague who is guided by the biblical precept that states that "In the beginning was the Word." For the former, historical change is but a subspecies of a more basic change taking place in the natural universe. His attempts to understand historical change eventually leads him back to the natural and physical world and his intellectual task becomes one of explaining human events and human motivations in terms that most closely resemble the formulations and descriptions used by natural scientists.

Within this perspective the human world makes its appearance as a mere fragment of the much larger and more enduring natural universe studied by astronomers and geologists. Within this light the whole of human history presents itself as but a minor and perhaps insignificant incident in the unending melee of material objects and natural forces. Historical change becomes here a subspecies of an ultimate and all-encompassing metabolic change that rules the material universe. That universe itself appears in this light as an ultimate instrument of equality in which everything moves toward confluence and dissolution of difference. It assumes the form of a gigantic, constantly churning stomach in which the rich variety of human persons, times and events is preordained to disappear without remainder. By contrast, a history that is guided by the biblical precept of Creation begins with a divine word and takes the form of a cosmic conversation. That history begins with an encounter between heaven and earth and between a mortal self and a divine other. Such a history begins when a first face lights up in the presence of another and when a first creative word opens a conversation and establishes a bond that lays the foundation for a human world. Let it be noted, incidentally, that a world that is governed and held together by the Word is a world that remains intrinsically mutable since everything within, all things material and spiritual, make their appearance within the light of an ongoing conversation.

I am reminded here of van den Berg's 1996 lectures at the University of Leuven in Belgium that ended on a very enigmatic and thought-provoking note. He cited the well-known first sentence of the gospel according to Saint John, "In the beginning was the Word (logos) and the Word was with God and the Word was God." He then added the following commentary: "The words: 'In the beginning' refer back to Genesis 1, 1 where we read: 'In the beginning God created heaven and earth.' It was thus in the beginning, but it did not stay that way. What remained was: 'The Word was with God.' What disappeared was: 'The Word was God.'"[1]

The Word that created a relationship between heaven and earth remained the Word that was with God in the sense that it remained divine while it suffused the whole of creation with wisdom and endowed it with purpose. But a word spoken or a word given is modified by the way it is heard or understood and as it begins to form an ongoing relationship. For a word to become effective in creation it must be *offered* as a gift and a

[1] J. H. van den Berg, *Geen Toeval* (Not an accident) (Pelckmans, Kok, Agora, Kampen, 1996), 155; my translation.

pledge, it must be made to pass a border and cross a threshold. It is only in this way can the Word can unite two separate domains into one conversational and metaphoric whole. When we stand within this perspective we see all historical events as eventually pointing back to the threshold that was crossed by the creative Word. It is this threshold that holds together all subsequent worlds, times and persons.

Van den Berg then adds the following remark: "Between the Logos that began to suffuse reality and God Himself a border or threshold was put into place. Before this border I come to a halt and remain standing: out of respect, but also with some regret."[2]

The ultimate gesture of the historian is not one of defiance before an obstacle he cannot overcome. It is a gesture of respect before a limit that has no place in a natural universe and that falls outside the scope of the natural sciences. Like a good guest he comes to a halt before a threshold that he cannot cross without the help of the host. He comes to a halt before a limit that holds host and guest together and that unites heaven and earth. Only a world that is united in this manner is endowed with history.

A Phenomenological Historiography

The author conceives of his work as phenomenological historiography. He wants to describe the significant events of a human world that we inhabit, in which we have a stake and that we call our own.

When he speaks of the "lived world" he speaks of a world that was born in dialogue and founded by the word. It is a world whose fundamental dynamism derives from a metaphoric rather than a metabolic activity.

---

[2] Ibid., 155.

Van den Berg has compared his manner of writing history to that of a portrait painter. He wants to create a vivid and accurate sketch of a particular era. He wants to engage that past era in a lively conversation and thereby shed light on his own life and circumstance. He seeks to remember the past because it is his best means to consciously live and understand the present. His historical interest can never be wholly separated from his commitment to a contemporary world in much the same way that the contemplation of a portrait can never be completely separated from a reflection on one's own life.

The author is particularly struck by the vivid interdependence of every aspect of a painting. He notes how even a minor change in hue or the slightest modification of a figure impacts dramatically on the appearance of the whole and changes its meaning. He writes:

> Instead of metabletics we might speak of historical phenomenology or phenomenological historiography. It is not a simple matter to describe phenomenology but an example might suffice here. Let us think of a painter in the process of creating a portrait. He looks alternatively at his model and then at his work in progress. Let us assume that he is not yet satisfied with his portrait, that something essential is still missing. He takes another good look at his model and then adds just a slight new touch to his portrait. It is this one touch that changes the entire work in progress. It puts all that he has painted thus far in a new light.
>
> It would be difficult to argue from a natural scientific point of view that the previously painted traits had been fundamentally and literally altered by the addition of the last stroke of the brush. But the phenomenologist is not a natural scientist. He proceeds in a different manner. From his perspective all the traits of the portrait have changed, have become other than they were before.[3]

[3] J. H. van den Berg, *Hooligans* (Nijkerk: Callenbach, 1989), 17–18; my translation.

Van den Berg then illustrates how this process of portrait painting is directly applicable to his manner of writing history. He tells us how he went about studying the French-German war of 1870–1871. He set out to paint a portrait of the era that would best represent the various and particular historical traits that he has been able to observe. Yet in the course of painting this portrait he becomes aware of something he had not seen before. He has learned that at the time of the war an important breakthrough occurred in the field of mathematics. This new information may at first glance appear to have little bearing on the theme of the portrait and the pursuits of the war. Yet the painter cannot neglect it since it has altered his perception and changed what he felt and knew about the era. A little later he learns that a new Parisian dress fashion, the so-called *cul de Paris*, had conquered Europe around the same time. It would seem far-fetched to suggest a meaningful relationship between the appearance of a new Parisian fashion and the outbreak and the conduct of a war. Yet, to the portraitist-historian this minor historical fact cannot be ignored since it formed part of that era and as such sheds its own inimical light on it. It should therefore find its place on the canvas of time. It goes without saying that a historical account thus considered remains always incomplete and demands forever to be repainted and rewritten. All that one may reasonably expect of a painter or of a phenomenological historian is an ever renewed effort to portray with a keen eye, with thought and charity, an ever-changing and ultimately mysterious human world.

Van den Berg's thesis about the mutability of material substances raises questions about the role he assigns to the natural sciences in our thinking about ourselves, our neighbors and our world. His general position is that these sciences have much to teach us about natural and human reality, provided that we do not take them collectively as an *ultimate* framework for our thought, or misuse them as a final guide to moral and civic actions. These sciences are helpful in our daily struggles with material nature but they cannot give us absolute or final answers as to how we should understand our world or how we should ultimately regard water, bread, wood or steel.

The historian, the psychologist or the sociologist working within a natural scientific framework has no choice but to understand the world he or she seeks to describe as *ultimately* revealed by geology, astronomy, physics, chemistry and biology. The metabletic historian, on the other hand, works within a very different horizon that is opened by the miracle of personal encounters. It places at the beginning of history a mutual revelation of self and other, of a host and a guest. It describes a human world that is founded on a covenant. This historical world is held together not merely by anonymous forces and natural laws. It is embraced and maintained by a first word that was pledged and by a conversation that was begun when heaven and earth were united and a first couple began to inhabit the earth. It will not end until there are no human beings left to hear the word and to maintain the conversation.

An inhabited world is made whole and coherent by a bond between a host and a guest. Only a world that recognizes and treasures that bond and that practices hospitality in all its spheres can form a fertile ground for science and technology. Only a world that makes the meeting of hearts and minds its central concern can give birth to works of art, poetry and history. Natural science permits us to understand the rainbow as a natural phenomenon. But natural science can prosper only in a world where it is still possible to see the rainbow as a pledge of fidelity and as a symbol conjoining heaven and earth.

About Natural Scientific Psychology

Van den Berg clarifies his understanding of the natural sciences with an example drawn from the field of experimental psychology. He asks us to imagine an experimental psychologist who seeks to understand depth perception and uses himself as a subject in an experiment designed to clarify certain aspects of

the phenomenon. He begins his work by distancing himself from his own ongoing lived experience of depth perception and by transforming it into an object of naturalistic observation. He experiments with his own bodily reactions to changing laboratory conditions and in that way gathers data which he then subjects to various thematic and mathematical analyses. He then terminates his investigation by writing a report on his experimental observations and conclusions.

The experimental psychologist is aware of the limitations that are inherent in his work. He knows that his descriptions and measurements should be seen within the context of a much larger, ongoing scientific enterprise. He is fully aware that he has not been able to study depth-perception from all possible naturalistic angles and under every conceivable material circumstance. He accepts these limitations while consoling himself with the thought that he has made a contribution to a scientific enterprise that some day in the far future will yield a nearly complete understanding of depth perception.

What van den Berg finds missing from this account is the psychologist's realization that his experimental study concerned only an objectified and naturalized depth perception and entirely ignored the question concerning the relationship between an objectified and a lived world. It thereby overlooked the need to integrate the objective study of a natural phenomenon within the larger context of an ongoing, actually lived human world. Van den Berg argues that the psychologist made use of this depth perception, that he inhabited it when he traveled from his home to his laboratory. He again depended upon it when he conducted his experiments and made his calculations. He remained anchored within it when he returned home from work and sat down with his family to dinner. It formed an inalienable part of him as he saw those around him and was seen by them.[4]

---

[4] Personal communication from J. H. van den Berg to Bernd Jager.

This failure to reintegrate an objectified and universalized world within the larger lived world from which it arose should not be attributed to an oversight of the psychologist or to a particular flaw in his experimental design. This oversight is inherent in the natural scientific quest itself and in the heuristic fiction that the lived body and the inhabited human world are first and foremost material things that belong to and are entirely enclosed within a natural scientific universe. This heuristic fiction reveals aspects of the human world that otherwise would remain hidden. But it also obscures other, essential dimensions of human life that cannot be revealed within the context of natural scientific narratives and practices.

To understand natural science as a revealing but limited heuristic fiction does not diminish its value, nor should it lessen our esteem for its brilliant accomplishments. Such an understanding upholds the integrity of a scientific narrative by clearly distinguishing it from religious, literary, philosophical or political narratives and preventing brilliant science from being transmogrified into bad poetry or destructive myth.

A scientific and naturalistic inquiry into human depth perception reveals on the one hand the intricate coordination of right-eye and left-eye vision. On the other, it reveals complex patterns of sensory interactions and synesthetic confluences that make possible the unitary revelation of a world. But what escapes this naturalistic inquiry is the mysterious interaction between a world revealed in objective observations and calculations and the vaster, wider, deeper and more profound lived world that supports it. Cultural coherence demands that we do not lose sight of the lived world as we explore the natural and objectified worlds that arise from it.

The universe of science can shed light on the world we actually inhabit only as long as we maintain a creative distance and difference between subject and object, between person and world, between the one who sees and the things seen. Thought and perception do not copy a natural world, they form together a dual, metaphoric unity of host and guest, of subject and world and thereby give access to a meaningful world.

The psychologist exploring his own depth perception should maintain a metaphoric distance and difference between a naturalized and objectified depth perception and the depth perception that forms part of his lived world and that enables him to objectify his world and undertake scientific studies. It is this latter depth perception that forms part of the inhabited domain and helps to create the platform from which it becomes possible to view a natural world and to undertake scientific studies.

The failure to maintain a metaphoric distance between self and other, heaven and earth or between a lived world and a natural universe translates into losing the means to integrate scientific findings within a larger religious or philosophical narrative. It shows itself in the example of the psychologist who thought his task was finished once he had succeeded in translating a psychological phenomenon into the language of biology, mathematics or physics.

We are reminded here of the apocryphal story told by Galileo's assistant Vincenzo Viviano about the discovery of the law of isochronism of the pendulum.[5] It tells how Galileo came to his important scientific insight while he sat in the cathedral of Pisa awaiting the celebration of the Mass. His attention was drawn to the swinging motion of the chandeliers as they were pulled down and then hoisted up again by the sacristan who was busy lighting the lamps. Galileo became fascinated by the swinging motion of the chandeliers and began to observe and time their oscillations. He measured this with the help of his own steady pulse and noted that the time needed for the completion of one complete oscillation was the same at the beginning of the process, when the swing was the fastest and

---

[5] Alexandre Koyre, *Etudes d'histoire de la pensee scientifique* (1966; reprint, Paris: Gallimard, 1973), 289–320.

the widest, and at the end when the chandeliers had almost returned to rest.

What interests us here is not the discovery of the law of isochronism itself but the circumstances that gave birth to it. Galileo's discovery depended on a creative leap of the imagination that permitted him to imagine the magnificent chandeliers of the cathedral as so many abstract pendulums cleaving an equally abstract and universal time and space. To be able to conduct his scientific observations the great scientist had to imagine a natural and temporal world that was in fact very different from the ceremonial and religious world in which he actually found himself. He had to imagine a natural universe in which there was no place for chandeliers, for masses, for sacred ceremonies, buildings or histories. In order to find access to what was to become a modern, natural scientific universe Galileo had to imagine a new space and time. He had to distance himself from the space and time of the cathedral and from the divine narrative and the sacred actions that gave it form and content. He also would have to take his leave from the civic, historical and political space of his hometown and even from the familial and amicable space and time in which he lived his personal and intimate life. In order to see the chandeliers as mere abstract pendulums whose movements were ordered solely by universal and natural law Galileo was required to embark on an audacious journey that for the time being would separate him from his church, his family and his hometown. Like the great seafaring explorers before him, Galileo journeyed to distant shores and discovered worlds that were stranger even than the ones discovered by Columbus and Cortez.

The discovery in the cathedral of Pisa should not be understood as an ultimate homecoming to an ultimate reality but as a journey to a far corner of the world that would not be complete without a homeward journey. It would not properly come to an end until the abstract pendulums of a universalized and naturalized world had found their proper place alongside the chandeliers of the cathedral. It would be vain and tendentious to characterize Galileo's new way of understanding the chandeliers as representing some absolute *progress* over his earlier, religious understanding. There is nothing in his discovery that would authorize us to see chandeliers exclusively as modified pendulums or to understand the cathedral as a mere prototype of Galileo's laboratory. Neither should it inspire us to *replace* an attitude of religious worship or of ritual celebration with an attitude of scientific inquiry. The cultural task imposed on us by Galileo's discovery is not one of substitution, of replacing one way of seeing and understanding with another, but of making place in our life for the cultivation of both attitudes and of creating a metaphoric whole out of what they reveal about our world. It is such integration that makes our world inhabitable.

The importance of Galileo's discovery lies in the fact that it enlarged the repertoire of the human imagination and that it created a different way of seeing and understanding our world. The challenge of every adventurous journey and of every great discovery is that of homecoming. It is this homecoming that marks the difference between provoking a destructive revolution and building a viable civilization.

It appears evident that in order to conceive of and observe a natural scientific universe one needs to inhabit a human world. The inhabited, metaphoric world constitutes the ultimate foundation on which rest all possible human observations and speculations. It forms the point of departure for all scientific, artistic and religious thought and practice. The heavens of astronomy and the mountains and seas of geology are accessible only to someone who is at home in the lived world, who has been cradled by a cultural life, who knows friendship and collegiality, who is upheld by divine, parental and conjugal love. These features of a strange and distant world can be explored only by someone standing on the shoulders of previous generations of explorers and thinkers and by someone offering his own shoulders for future generations to stand on.

The mountains of geology and the stars of astronomy become meaningful only when the abstract universalized world of which they form a part enters into a vivid dialogue with the lived world. These abstract features of a geological or astronomical landscape remain stillborn until the time that they find their assigned place alongside the mountains explored by hikers and cultivated by farmers and the stars admired by poets and lovers.

The Lived World and the Natural Universe

In their insightful study of van den Berg's work, Vandereycken and De Visscher have paid close attention to the recurring theme of the modern destructive tendency in the human sciences to constantly elevate a second, abstract "underlying" reality over the primary reality unfolding before our very eyes. They draw a clear distinction between a lived or inhabited world and an objectified natural universe. The first of these is in constant flux and is governed by a primary structure that mediates between self and other and between person and world. The second is governed by a natural laws that governs infinitely repeated and fundamentally inalterable conditions.[6]

To describe the lived world we must enter into a poetic or painterly perspective that opens upon a constantly changing physiognomic world of mutating relationships that are marked by spontaneity, surprise and discontinuity. To find access to a natural universe we must let ourselves be guided by an objectified or secondary perspective that leads us past the distractions of ephemeral and constantly changing everyday realities. The universe we discover in this manner accords no

---

[6] W. Vandereycken and J. De Visscher, *Metabletische perspectieven: Beschouwingen rondom het werk van J. H. van den Berg* (Metabletic perspectives: Reflections on the world of J. H. van den Berg) (Belgium: Acco, Leuven, 1995).

privilege to persons or to faces. It makes no ontological distinctions that have their roots in dialogue; it leaves no place for thresholds and repels all attempts at inhabitation.

This understanding, which plays such a large role in van den Berg's writings, also plays a central role in the work of his teacher and colleague Gaston Bachelard. We should recall that Bachelard described the birth of the modern sciences as a progressive depoeticization of daily life and as a gradual ascendancy of the concept over the image. He understood his poetics as moving in a direction opposite to that of scientific abstraction and naturalization. The first lines in the introduction to his *The Poetics of Space* read: "A philosopher who has evolved his entire thinking from the fundamental themes of philosophy of science, and followed the main line of active, growing rationalism of contemporary science as closely as he could, must forget his learning and break with all habits of philosophical research, if he wants to study problems posed by the poetic imagination."[7] This message was not lost on van den Berg at a time when he sought to distance himself from the naturalizing and geometrizing tendencies in psychiatry and psychology and sought to find his way back to a lived or inhabited world of personal relations. The same road that led Bachelard to explore the poetic imagination would lead van den Berg to develop his physiognomic and phenomenological conception of history. In his psychiatric practice it would steer him away from all attempts to anchor his observations in a secondary world of material realities and encourage him to explore the lived world of his patients.

From early on in his career van den Berg resisted the ingrained cultural habit of the day of disregarding ordinary lived reality in favor of an underlying, presumably more profound

---

[7] G. Bachelard, *The Poetics of Space*, trans. Maria Jolas (1958; reprint, New York: The Orion Press, 1964), xi.

and revealing second reality. He tells the anecdote of how as a medical intern he was introduced to play therapy. He observed a therapist interacting with a lively young boy who mixed sand and water in a bucket and then gleefully plunged his hands and arms into the mud. In any person not versed in the art of psychoanalytic interpretation this scene would evoke memories of times spent at the beach building sand castles and of being temporarily released from the parental edicts about cleanliness and propriety. Van den Berg was therefore shocked when afterwards he heard the therapist interpret the scene in terms of an "anal fixation" and let it be known that the child had played not so much with sand and water as with his own feces.[8] It seems that from very early on in his career the author wanted to elaborate a psychology that would not abandon the lived world in search of depoeticized abstractions and that would remain faithful to the infinite riches of everyday life.

If we try to further sharpen our focus on what distinguishes a natural universe from a humanly inhabited world we must pay attention to the manner in which they come into being and begin to form a coherent whole. A natural universe comes into being as the result of an *accidental* process, that is, of an accidental "falling together" (*ad-cadere*) of its component parts. The natural history of the universe is the history of a fall (*casus*); its dynamics is one of events running their natural course from high to low, from difference and diversity to literal and material unity, from flames to ashes. The coherence and unity of a universe refer back to chance events that are the result of an accident. By contrast, the unity and coherence of a lived world can only be understood in terms of a miraculous encounter in which one person becomes present to another and in which together they begin to form a metaphoric, dual unity. There where the universe demands to be understood in terms of chance events

[8] Van den Berg, *Geen Toeval*, 14–15.

that "fall together," there the lived world requires to be understood in terms of creative acts that actively and consciously "bring together" a human world. A universe "happens," but a human world can come into being only by being brought together, that is, by being created and by being assiduously cultivated.

The history of the lived world begins with a miraculous and personal encounter. We find access to this world by responding to an invitation; it is founded on a covenant, on a pledge of love and friendship, on a word given and a word received. We enter the lived world as we enter a house, by paying our respects to a threshold, by honoring a pledge and by freely entering into a reciprocal relationship of host and guest, child and parent, husband and wife. We enter it as friends, as neighbors, colleagues and fellow citizens. In doing so we contribute to its coherence and help give it meaning. We leave that world by giving our blessings to those we leave behind and by transferring our task to succeeding generations. We abandon it by surrendering it to "the elements" and letting it fall apart.

Van den Berg stresses the miraculous nature of human and divine encounters and implies thereby the miraculous nature of the lived world itself. He describes two such encounters in *The Changing Nature of Man*. The first of his essays concerns an anecdote from Freud's *Three Contributions*. It tells the story of a small boy staying the weekend with his aunt and becoming scared at night in his unfamiliar surroundings. The child cries for help, "Aunt, please say something. I am scared; it is so dark." The aunt asks a bit teasingly how her talking could lift the darkness, and the boy answers with the unforgettable line: "Aunt, when you talk it gets light."[9]

To the child the familiar voice of the aunt means light in the

---

[9] J. H. van den Berg, *The Changing Nature of Man* (New York: Delta Books, 1961), 195.

darkness. It restores a covenant; it re-establishes in a miraculous manner the lived order that supports the human world and makes it inhabitable. The order of the lived world should not be confused with that of a natural universe. The order of a natural universe can be grasped in the form of a material or logical principle or law that once it is understood grants us mastery over a natural domain. But the order that brings light to the dark room issues from a restored pledge and a renewed offer of hospitality. That light and that order emanate from a dependable and loving relationship. It is in the light and the ordered circumstance of that relationship that the child is able to find his place in the world. It is from that place and while standing in that light that he may explore the miraculous nature of the lived world or, on the other hand, learn to investigate the natural order of a material universe.

Van den Berg derives a second anecdote from André Gide's autobiography *Si le grain ne meurt* (If it die). It tells of one of the author's earliest memories of a splendid walk in the countryside with his beloved nurse. Gide recalls that on that particular day his nurse appeared radiant with happiness and he had asked her what made her feel that way. She answered innocently, "Nothing in particular. But isn't the weather gorgeous?" Gide recalls that when he heard those words, "the whole valley became filled with love and happiness." It was as if the smile of the nurse granted the boy new access to a landscape that up to that point had been perhaps no more than an indifferent expanse. The smile and the words of the nurse miraculously transformed it into a valley filled with promise and delight. The flowers became more colorful, the shadows grew suddenly deeper, the blue vault of the sky became more impressive and the sun more radiant. This metamorphosis came about through nothing more substantial than a few words, an eloquent gesture and a smile. But these few words and that smile gave new life to an interpersonal bond capable of ordering the world anew and making it more available and inhabitable.

The child crying out to his aunt in the darkness sought reassurance of a relationship that for a moment he feared might be lost. He tapped, as it were, the ground to test its solidity and to make sure that it could bear the weight of his existence. The child of the second anecdote perhaps felt somewhat estranged from his nurse because of a happiness in which he did not share. He also sought reassurance about the firmness of the ground underfoot and the solidity of a personal bond. Once she shared that happiness, once that bond was re-established and that foundation secured, an indifferent earthly expanse became miraculously transformed into a promised land that awaited his exploration. The landscape *offered* itself to be inhabited so that the hills invited the child to skip or roll down its slopes, the trees bade him to climb up their branches and the butterflies asked to be observed and chased. All these possibilities of the landscape were directly linked to the hospitable presence of a near-dwelling or *neighboring* person. If the nursemaid had suddenly fainted or for some other reason broken off all further contact with the boy the invitation and the promise would have been withdrawn from the landscape. Its hills would have stopped all incitement to running and rolling and the butterflies would have disappeared beyond the reach of the boy. The color and the golden light would have drained away from the world. And if the aunt had not answered the child's call, all comfort and warmth would have disappeared from the child's bedroom. A humanly inhabited, lived world finds its prototype in a welcoming home. It unites and holds separate two neighboring domains, that of the self and the other, and that of an inside and an outside. The dynamism of this world takes the form of an unceasing, metaphoric exchange between these domains that is governed by a threshold. This threshold makes possible the cultivation of both an inside and an outside. It makes possible the cultivation of an inside world of friendship, of family and religious bonds, of citizenship and collegiality. But it also makes possible the systematic exploration of an inhospitable outside world, of a natural universe or a no man's land, where things accidentally *fall together* rather than being hospitably *brought together*. A truly human world is born only there when the knowledge of how things fall together (*ad cadere*) becomes fully integrated within a whole that we experience as having been brought together.

Sacred and Profane Transubstantiation

Van den Berg concludes his essay with an enigmatic reference to the Christian doctrine of transubstantiation. He restates his belief in the close relationship between the French Revolution and Count Rumford's discovery of the law of the conservation of energy. He makes it clear, however, that he does not think about that relationship in terms of a materialistic theory of cause and effect. Neither does he accept a romantic and individualistic theory that would make Count Rumsford's individual genius solely responsible for his scientific discovery. He sees both the Revolution and the simultaneous advances in physics as announcing and exemplifying, each in their own way, a more fundamental and more general change in the relationship between heaven and earth, mortals and immortals, divinity and humanity. This changed relationship affected not only hearts and minds but also changed the nature of the material world insofar as it was revealed in the light of that relationship. It was Rumsford's genius that first detected this changed nature while he supervised the manufacture of cannons in München.

We should remind ourselves that the modern rejection or negation of an ongoing conversational relationship between heaven and earth constitutes by itself a metabletic change that necessarily affects our understanding and perception of the natural world. Van den Berg understood the fundamental change that took place in the Western world around the time of the French Revolution as a *profane transubstantiation*, that is, as a miraculous change that took place "before or outside the temple" (*pro-fanum*). He implies that this miraculous change bears a certain resemblance to the miracle of a divine encounter inside the temple.

The miracle within the temple celebrates the creation of the human world as a coming and bringing together of an inhabitable realm that makes place for both divinity and humanity, for both heaven and earth. It reminds the faithful that the human world was born in a festive encounter in which a divine being came down to earth, broke bread with mere mortals, entered into a new alliance and opened a conversation in the light of which all things mortal and immortal were transformed and endowed with new purpose.

We might step beyond what van den Berg would be willing to assert when we interpret profane transubstantiation as a historical turning away from the temple that led to the discovery of a modern universe born of accident and governed exclusively by natural law. Left to its own devices and completely cut off from what takes place in the temple, such a profane transubstantiation would eventually assume the form of a purely metabolic process in which the human world would be made to disappear.

Divine or sacred transubstantiation, on the other hand, would refer to a homecoming to the human world of metaphoric change and to a specifically human and divine *bringing together* that builds an inhabitable cosmos. Metabletic history appears in this context as an attempt to safeguard the integrity of both a sacred and a profane transubstantiation taking place within and without the temple. Its essential task would be that of drawing both together within the metaphoric whole of an inhabited place and a lived world.

History's primary task becomes here one of reestablishing forgotten links, not only between the past and the present, but also between a profane and a sacred realm. Its main task would be that of mapping and describing the various ways in which a past and a present, an inside and an outside, and a profane and sacred time and space can be brought together to form an inhabitable, metaphoric whole.

Bernd Jager
Université du Québec à Montréal

Translator's Introduction

Towards the end of the Middle Ages a growing number of clerics and scientists became aware of the need to thoroughly rethink ecclesiastical doctrines concerning the truth of natural scientific investigations. They realized that there was an urgent need to secure a legitimate place for a mundane, unrevealed and natural scientific truth alongside, and in dialogue with, the sacred, revealed and theological truth of the Church. Most importantly, they became convinced that the theological exploration of biblical truths, of dogmas and articles of faith, required a different approach and took place within a different intellectual horizon than did the explorations of mathematical and natural scientific truths. They saw the one approach as predicated on mathematical necessity and practical and material proof while the other approach depended on divine revelation and grace.

It was in this way that a doctrine of two separate truths gradually developed with each addressing a particular and distinct realm of inquiry. One realm of truth was revealed by Holy Scripture and was based on faith, while the other concerned a nonrevealed secular realm, disclosed authoritatively by logic, secular observation and experimentation. Presented in this manner each of these realms could be thought of as governed by distinctive rules, procedures and laws and demanding a different interpretive attitude and methodology.

Up to that time the Church had held to the position that there could be only one single, incontrovertible, revealed truth that governed every aspect of human life on earth and that was binding for both theological and natural scientific explorations. No doubt the Church feared the division and the eventual degradation of the one ultimate and divine truth of which it was the unique guardian. Moreover, it may not have relished the thought of a rival group of worldly clerics and scientists establishing itself alongside the hierarchy of the church and claiming authority over a realm that up to that time had been under the direct authority of revealed religion. It was no doubt for these and similar reasons that the theory of the dual truths was formally rejected by the Fifth Lateran Synod of 1513.

This official doctrinal pronouncement of the synod did not bring the controversy to an end, however. As mathematicians and natural scientists continued their explorations of the natural world in a nontheological direction they found themselves in growing conflict with biblical and revealed truth. This course of events led to a tragic collision between the two sides during the trials of Galileo in the seventeenth century.

It is well known how the then emerging technology of lens-grinding in Holland had enabled Galileo to detect craters and other irregularities on the surface of the moon. This new discovery made clear that the moon and other heavenly bodies displayed irregularities and imperfections that up to that time had been exclusively associated with terrestrial realities. This observation ran counter to astronomical conceptions established in antiquity that presented the earth and the celestial realm as essentially different entities that obeyed different laws and required a distinctly different cultural approach for their observation and understanding. At the time when the Church's position hardened into a rigorous monism of revealed truth, the seeds were sown for an equally dogmatic monism that would later proclaim that a single natural scientific approach would suffice for the study of both heaven and earth. In the human sciences this would eventually lead to a dogmatic belief that the same principles and methods used in the study of astronomical or biological objects would suffice to reveal the whole of human reality.

It was in this way that the dogmatic monism of the Church that sought to circumscribe natural scientific investigations transformed itself over time into the dogmatic monism of various materialistic and scientistic ideologies that sought to establish themselves as the sole legitimate sources of light and truth about the human condition.

Van den Berg clearly accepts neither the ancient *sola fides* nor the modern *sola scientia*, and the essay that follows can perhaps best be understood as a reflection on the relationship between these two very different visions of the world and as a contribution to their fruitful coexistence. He concerns imself very little with the proposition of a *sola fides* but he places himself squarely in opposition to the scientistic ideologies of *sola scientia* that have been the cause of so much suffering and destruction in the twentieth century. He clearly rejects the metabolic frame of mind that would collapse the sacred into the profane or the profane into the sacred, or that would make the past disappear into the present or the other into the self. He objects to a frame of mind that would wish to suspend all exchanges between neighboring worlds and silence the dialogue between various traditions, disciplines so as to restrict our cultural life to the observation and mastery of an objectified natural scientific universe.

It is for this reason that he objects to a psychological frame of mind that ignores the all-important difference between lived perception and an objectified and universalized representation of that perception. Van den Berg insists that the experimental psychologist's work of abstraction and objectification must be supplemented and completed by a work of cultural integration.

Such integration seeks a proper place for a particular scientific insight or discovery within the larger world of human sensibilities, purposes and understandings.

The experimental psychologist's assumption that his perception should be exclusively understood in terms derived from biology and natural science can best be approached as a kind of fiction that permits a particular, limited disclosure of the human world. Such heuristic fictions sustain the physicist or the psychologists in their particular pursuits that, if successful, shed a particular light upon our world.

To understand natural or human science as inspired by and based upon such revealing fictions does not denigrate it or diminish its vital importance to our lives. It only points to the fact that natural science cannot give us a complete or self-sufficient account of human existence. It proposes that what natural science tells us about ourselves and our world needs to be complemented by the revelations and elucidations of other disciplines and traditions that use other narratives to reveal our world.

The work of understanding and making sense of our world cannot be achieved solely by a process of natural scientific objectification and calculation. It needs to be complemented by a labor of integration that prepares a place for every new insight into the natural universe within a more encompassing human world that is governed by laws and formed by purposes and ideals that fall strictly outside the limits of a natural universe. It is in this larger, more varied and complex human world that we must live our lives, build our homes, raise our families and construct cohesive communities.

If we think of natural science as guided by an extremely useful and fruitful fiction we are still confronted by the problem of gracefully integrating that fiction within the larger narratives that give sense and purpose to our lives and that permit us to feel at home in the world.

The failure to integrate natural scientific narratives within a more encompassing narrative bears certain inevitable consequences. One of these is that while science and technology may for a while continue to prosper, all the other arts, disciplines and practices begin to fade and to disintegrate as they attempt to imitate scientific rationality by adopting naturalistic and technological objectification as their starting point. This deformation and self-mutilation prevents these arts, practices and disciplines from shedding their own particular light on the human condition and forming a necessary cultural complement to science and technology.

To counter this trend it is necessary to remind ourselves that natural scientific discoveries are not monolithic achievements that take place in a cultural vacuum. Van den Berg's essay shows us the limits that are inherent in the revealing fictions of objectification and naturalization. Unchallenged by other cultural practices and disciplines, natural science promotes a vision of the human world as but a tiny and ultimately insignificant part of a natural universe set in motion by natural forces and ruled exclusively by the logic of physical necessity. In the grip of this narrative we are in growing danger of experiencing our own bodies as a kind of natural assembly of separate biological organs and functions that should be approached and evaluated exclusively from within the perspectives of medicine, biology and biochemistry. It is in this way that human life comes to be regarded as in essence a natural, biological process governed by the same laws that rule the rest of a natural universe.

The heuristic fictions of biology and physics clearly have their value and their place within the limits of the scientific theories and practices they serve. But once they are adopted as a master narrative that guides our personal and communal life they betray us and lead us to ruin. They serve us so well when we attempt to objectify and naturalize our world, but they fail us utterly when we attempt to build a human world by forming friendships, by establishing families and building neighborhoods and cities.

To protect us from the ruinous and perverse use of the natural sciences it is useful to be reminded that scientific discoveries do not occur within a cultural vacuum. In the pages that follow, van den Berg provides us with a stimulating account of the broader cultural world that forms the background against which the important scientific discoveries of the late eighteenth and the beginning of the nineteenth century took place. His essay shows the close alliance between scientific and political revolutions.

Modern chemistry textbooks still use the notations and some of the fundamental concepts introduced by the founder of modern chemistry, Antoine Laurent Lavoisier, at the time of the French Revolution. Lavoisier's work brought an end to the ancient Greek conceptions of a material world in terms of the four basic elements of Earth, Water, Air and Fire. These antique conceptions formed part of a coherent tale of the world that ordered the world so as to make it inhabitable and to find within it a just and fitting place for all the things and beings that it contained. These ancient elements formed part of a narrative that aimed to reveal not a natural universe but a lived world.

It is clear that there is no a place in a modern physical universe for the ancient elements, but they continue nevertheless to shed their revealing light on the world in which we live our daily lives. They formed part of a prescientific world worldview in which water and air continue to form part of a specific landscape or locale. In that view, the water of the Seine is not exactly the same substance that makes up the Danube, and the air of Paris is never exactly that of the Provence or of the Alps.

After Lavoisier, all the chemical elements became detached from the human habitat. Water and salt became substances that no longer maintained intrinsic links to a human world; they remained self-same compounds no matter where and under what circumstances they were encountered. It is in that sense that Lavoisier's thought led us away from concretely experienced water, from the water of bathing, drinking, fishing and boating and presented us with an entirely new experience of water as an abstract and universal compound made up of two different elements.

Van den Berg points out how Lavoisier's vision of the equality and universality of chemical elements and compounds occurred around the same time that the French Revolution abolished difference based on birth and class, proclaiming the equality of all citizens. Among the three great principles advocated by that Revolution, those of freedom, equality and brotherhood, it is certainly the one concerning equality that defined the heart of the Revolution. The French Revolution, which was at the same time a revolution of the entire Western World, began in 1789. Lavoisier's revolutionary book on chemistry appeared in that same year, so that we may think of that year as having given birth to both a political and a scientific revolution.

It is possible to see this twin birth as a mere coincidence and consequently reject the idea of any intrinsic relationship between the two. But it is also reasonable to see both revolutions as growing out of a converging underlying attitude of abstraction and geometrization. In that case we must let go of any cherished parochial notions about an absolute wall of separation between political and intellectual ideas and pursuits and look for more fundamental attitudes and beliefs that form a link between them.

In what follows Van den Berg will maintain that the universalism of modern natural science grew out of a larger climate of thought, attitudes and feelings that marked a particular historical era. It places the objectivity and universalism embodied in the stance of the natural scientist within the larger context of the equality and universalism promoted by revolutionary utopias. The merit of this approach is that it seeks to reintegrate a particular stance developed in the natural sciences within the larger lived and inhabited world from which it arose. That larger world includes political upheavals as well as artistic, literary, religious and philosophical activities.

To understand natural science as forming a part of this larger cultural complex does not denigrate it or interfere with its integrity or autonomy. On the contrary, we can develop a proper appreciation and understanding of natural science only by finding a just and proper place for it within the diverse whole of cultural endeavors. It also will help us to make a clear distinction between the practices of natural sciences and the very different cultural practice that seeks to integrate scientific viewpoints and discoveries within a larger cultural landscape.

We should keep this distinction in mind as we read van den Berg's account of the historical and cultural circumstances that surrounded the discovery of the first two laws of thermodynamics. This account is clearly not intended to add to our *scientific* understanding of these laws, nor does it seek to contribute to a history of scientific progress viewed from the perspective of the sciences. Its aim is rather to draw the world of science back into the larger sphere of cultural life from which it sprang and to which it must return if it is to make its proper contribution to our lived world.

How, then, are we to understand the shift in human understanding that occurred toward the end of the eighteenth century and that gave rise, all at the same time, to natural scientific discoveries and political revolutions? Van den Berg describes it in enigmatic terms. He writes: "Rumford lived at a time when metal underwent transubstantiation." The author tells us about his initial hesitation to use the term "transubstantiation" to describe a historical change. He was reluctant to describe a historical change in terms of a theological concept that evoked the central mystery of the Christian faith. Theologically the term refers to the conversion of bread and wine in the Eucharist into the bodily, personal presence of Christ.

No doubt van den Berg chose the term initially because he wanted to refer to a change that occurs in history but that could not be interpreted in terms of a series of interlocking causes and effects. He wanted to specify that the change that occurred in metal at the time of Rumford could not be explained by changes in the psychological or sociological climate of the time, or attri-buted to political or economic causes, or subsumed under the general rubric of scientific progress or of technological advances. He proposed instead that the changes that occurred in Rumford's metal manifested themselves at the same time and in different ways in political upheavals, in dietary habits, in legal pro-ceedings, in scientific discoveries and in economic innovations. Yet none of these cultural realms should be thought of as being the *cause* of the changes that occurred in the other domains of cultural life.

Van den Berg insists that the historical change that ushered in the era in which the laws of thermodynamics could be discovered cannot itself be fully analyzed in terms of material causes and their effects but must be understood in terms of a fundamental shift in the relationship between human beings and their world. This shift belongs to the realm of personal relations rather than to that of material or mechanical inter-actions. Suddenly the world appears in a different light, shows a different face and permits us to see something that we had not been able to see before. Van den Berg's analysis makes us suspect that behind the silent spectacle of changing historical facts and circumstances there remains to be discovered a different history of a changing relationship between heaven and earth that is reflected in all the different manners in which we inhabit the earth. Transubstantiation can then be thought of as referring to a new and changed relationship between heaven and earth that is reflected in all manner of personal and material relations. We then see it as a metamorphosis in the light of which the whole of human reality appears in a new light.

If we accept this interpretation we are confronted by two types of history, or by a history that manifests itself on two different levels. On the one level we meet with a history that records the various mutations of sense and substance in a world that is in perpetual motion. It offers us the chance to privilege one particular domain and then consider it to be the cause of change in all the others. We might choose to privilege economic or political activity that way, or elect scientific progress or the onward march of literary fashions or of philosophical ideas. We may think in terms of outstanding personalities or creative geniuses and think of these as the ultimate sources of historical change. With the term “transubstantiation,” van den Berg chooses a different path and evokes the ancient image of a human encounter with the divine as the principle and the source of both personal and historical transformation.

In concrete historical terms, van den Berg uses the phase “transubstantiation of metal” to refer to a change that occurred at the time of Rumford. It transformed the old metal that had been a local, handcrafted product and an inalienable aspect of a particular landscape into a universal chemical compound that belonged entirely to the abstract universe of the natural sciences and its progeny: modern labor and industry.

This transformation can be thought of as an indifferent rearrangement within a material world or it can be understood as revealing a fundamental change in an ongoing dialogue between heaven and earth and as a sign of changed human and material relations.

Van den Berg’s introduction of the term “transubstantiation” can be read as the author’s attempt to restore to the silent and universal world of material causes and their effects the sound and sense of dialogue and encounter.

We are reminded here of Merleau-Ponty, who in his last book made a similar, seemingly secular use of the same theological concept. He used it to describe what he saw as the miraculous transformation that takes place when a painter changes something he sees into something painted. He writes: “It is by lending his body to the world that the artist changes the world into paintings.” He further notes that: “To understand these transubstantiations we must go back to the working, actual (human) body — not the body as a chunk of space or a bundle of functions."[1] The miracle of transubstantiation refers here to an active embodied engagement and encounter with the world that seeks to reveal and to be revealed.

As we saw earlier, van den Berg has compared his manner of writing history to the painting of a portrait. To succeed in this task the historian, too, must lend his embodied existence to the task so that a full and rich portrait of an era can come into being. Where the objectifying natural scientist must practice withdrawal, must remove from what he wishes to see any trace of his own particular, embodied existence, the portrait painter and the metabletic historian must engage with and fully inhabit what he seeks to understand.

To understand painting and the writing of history in the manner of van den Berg, we must reach beyond the merely technical world of oil and pigment and the historical world of material facts and their arrangements to discover beyond the world disclosed by factual and naturalistic observations a very different, primary world. That primary world is a place of encounter, dialogue and transubstantiation before it becomes a world of mathematical and natural scientific abstractions.

It is in the same way that the human body is a place of miraculous transformations, and indeed the original abode of transubstantiation, before it can take on the form of a biological or material thing and become the object of natural scientific inquiry. Merleau-Ponty observes

> We can speak of a human body when, between the seeing and the seen, between touching and the touched, between one eye and the other, between hand and hand, a blending of some sort takes place — when the spark is lit between sensing and the

[1] M. Merleau-Ponty, *The Primacy of Perception* (Evanston, Ill.: Northwestern University Press, 1964), 162.

> sensible, lighting the fire that will not stop burning until some accident of the body will undo what no accident would have sufficed to do.[2]

A human body is born when a mysterious spark is struck and a fire is lit that illuminates the world. Transubstantiation refers here to a luminous combustion or a sacred fire that will burn until the moment when an accident of the body undoes *what no accident by itself would have been able to bring about.*

The human body cannot be fully understood when we view it solely as a coincidence of accidental factors. And neither can we grasp history in general and the history of science in particular in that manner. Understanding history ultimately requires making reference to a miraculous change belonging to a sphere beyond the reach of factual analysis and naturalistic investigations. This miraculous change occurs at the point of interlacing perspectives and at the moment of an encounter between neighboring worlds.

In the essay that follows, Van den Berg does not elaborate on the nature of this miraculous or "metabletic" change that transformed the material world at the time of the French Revolution and prepared it for Rumford's discovery of the laws of thermodynamics. He insists, however, that this fateful change cannot be fully understood from within the framework of the natural sciences and requires a return to the inhabited world in which we live our familial, civic, artistic and religious life. To understand historical change we ultimately need to reach beyond a purely naturalistic framework and confront the miracle of transubstantiation in all the human and divine ways in which it manifests itself.

Bernd Jager
Université de Québec à Montréal

[2] Ibid., 163–64.

Two Laws

Introduction

In this essay we will examine the two main laws of thermodynamics, the natural science that studies the relationship between heat and mechanical power. The Greek word *thermos* refers to heat and *dynamis* to strength might or power.

The eighteenth century saw the development of the first functional steam engines. For example, the famous mechanical water pump invented by Thomas Newcomen in 1711 was used for clearing water out of mine shafts. These new engines raised unprecedented questions about the nature of heat and power and their precise relationship. What, for example, was the connection between the heat released by burning wood or coal and the resultant work performed by these new inventions?

The nineteenth century witnessed the development of the coal-burning locomotive. The very sight of these enormous engines scaling the sides of mountains, passing through tunnels and roaring across bridges left an indelible impression on all who saw them for the first time. It also set in motion new ways of thinking about the relationship between heat and energy (*thermos*) and the work performed as a result (*dynamis*).

The first law of thermodynamics states that there is no loss of energy when heat is converted into mechanical power. Both are considered to be different forms of the same energy. This is not only true for power generated by heat in the steam engine, but also counts for heat or power generated by electrical or chemical processes, or by those created by wind currents or waterfalls as well. In all cases, the first law states, we are dealing with different forms of the same energy. The first law of thermodynamics is called "the law of the conservation of energy." It is a law that is valid for the whole of the natural universe.

The second main law of thermodynamics holds that the spontaneous flow of heat from hot to cold bodies is reversible only with the expenditure of mechanical or other nonthermal energy. In other words, the transformation from heat energy to mechanical energy can occur only when heat flows from a warmer body to a cooler body. This means that a steam engine had to have both a source of heat and a cooling element or condenser. In the early steam engines the outside atmosphere surrounding the locomotive provided the needed cooling effect, but later models had a condenser built into the engine itself.

The second law also holds that at each particular instance in which heat energy is converted into mechanical energy only a part of the generated heat is actually converted into work. The rest of that heat energy is not actually lost — the first law would otherwise be contradicted — but becomes scattered in the process. The official term for this scattering of heat is *dissipation*, a word derived from the Latin *dissipare*, which refers to a scattering, frittering or wasting away. Bearing in mind that the first law is commonly called the law of conservation of energy, the second could be termed the law of the loss of energy, with loss referring to energy that is scattered rather than destroyed.

The first steam locomotives amply demonstrated the principle of scattering or dissipation of heat predicted by the second law. The amount of heat successfully converted by these early machines into useful labor amounted to no more than 3 percent of the total heat generated by the burning of wood or coal. Nearly 97 percent of the remaining heat literally disappeared into thin air, not to mention the energy consumed by the friction of the wheels on the rails and other, smaller energy-consuming processes. Even later steam engines, insulated in sealed spaces and using ingeniously engineered safeguards, still lost a considerable part of their heat through dissipation.

This preliminary sketch of the two laws of thermodynamics suffices for the scope of our study, which is to place these advances in natural science within a larger cultural and historical perspective. Only a very general understanding of the laws and their implications is needed for this purpose. It was the American Rumford and the German Mayer who made the most important contributions to the development of the first main law of thermodynamics. In the next two chapters we will examine the specific methods, outcomes and implications of their scientific experiments. The most important contributions to the development of the second main law came from the Frenchman, Carnot, the German, Clausius, and the Scot, Thomson, who later became known by the name of Lord Kelvin. We will refer in more detail to these contributors as they become relevant to our exploration.

We make use of the term *metabletics* to characterize our historical approach. This term derives from the Greek verb *metaballoo*, which refers to plowing or turning the earth. It also refers to changing one's way of life, one's mind or one's purpose. In this study, the term *metabletics* refers to the study of radical changes in the course of history. These changes often become visible simultaneously in different areas of cultural and historical life. The purpose of this study is to place the discovery of the two laws of thermodynamics within the broader context of contemporaneous cultural and historical events. By placing them within this context, new and unsuspected aspects of these laws will come to light.

We cannot proceed with our inquiry unless we make use of a general concept that refers to the material nature of things. The concept of "matter" as it is currently understood, however, appears inadequate for our use because it has become so completely entrenched within a natural scientific context. It therefore seems preferable to use the concept of "substance" since it retains links with premodern ways of thinking. The concept was in use in antiquity even before Aristotle adopted it as a key element in his thought. *Sub-stance* literally means "that which lies underneath" something and thereby sustains and supports it. It refers to what a thing is in its essence, or to the essential content of a thing. It refers to the carrier of all the accidental qualities that collectively define a thing.

When the great discoveries in thermodynamics were made a major change took place in the essential nature or substance of all things. It was this change that made the dramatic developments in the fields of science and technology possible and opened a path for the development of thermodynamics.

Chapter I

The First Main Law: Count Rumford

Rumford was a singular and complex character who was not easily understood, even by those who knew him well. His singularity of character manifested itself in the way he discovered the first law of thermodynamics, and it is important to outline the main events of his life for this reason.

Rumford was born in the small town of Concord, formerly named Rumford, in the Commonwealth of Massachusetts in 1753. Upon his birth he was given the name Benjamin Thompson. In 1772 he married a rich widow, thirteen years his senior, who gave birth to a daughter. During the American War of Independence he sided with the British, enlisted in the army, and was promoted to lieutenant colonel even though he never saw actual combat during his brief career. When the war came to a sudden end he was forced to immigrate to England.

He left his wife behind in the United States and never saw her again. He does not seem to have missed her terribly, and there is no sign that bachelorhood disagreed with him in the least.

In England he received the title "Sir" as a first step toward his inclusion into the ranks of British nobility. After his promotion to colonel, he sought a wider scope for his activities and obtained permission to travel to the Continent. He ended up in Bavaria, where he entered into the services of the elector as an administrator. Moving as he did among the most prominent circles of the gentry and the nobility, he soon attracted the favorable attention of the court. Accounts from that time note that he was generally admired for his excellent manners and dynamic energy. It seems that Bavarian society was nearly unanimous in its acclaim of his intelligence, integrity and generous spirit. These qualities, together with his military experience, made for his rapid advancement within the ranks of the Bavarian government. After a phenomenally short time, Sir Benjamin was appointed minister of war and minister of the police. He was finally declared viceroy of Bavaria and enjoyed an unusually high level of public and collegial esteem. He was one of the first ministers of government to launch campaigns against poverty, disease and crime. Revered as a public benefactor and military reformer, he improved soldiers' conditions and increased military effectiveness. In addition, he sponsored the first "free meals" for the poor. We will explore this particular topic more fully later.

Once he was bedridden with a serious illness. On that occasion a procession of hundreds of local townsfolk filed past his house, each bearing a lit candle which they brought to the local cathedral in order to hold a vigil for his swift recovery. At the sight of this crowd filing past his window Rumford exclaimed: "And they do all this for a Protestant!" He received further official honors around that time. The Prince-elector bestowed the title of Count of the Holy Roman Empire upon him. It was upon the reception of this hereditary title of nobility that the former Benjamin Thompson adopted the name Rumford, after the town of his birth. He was known from then on as Count Rumford.

Rumford in Paris

Rumford spent his last years in Paris, where his tombstone can be found engraved with laudatory inscriptions. In Paris he fell in love with Madame Lavoisier, the widow of Antoine Laurent Lavoisier, who is generally recognized as the father of modern chemistry. Antoine Lavoisier had lost his life earlier under the guillotine. During his brief trial the judge Dumas had voiced the opinion that the Revolution had no need of chemists, and condemned him to death.

The widow of Lavoisier resembled Rumford's first wife in several respects, not the least of which being that she possessed great wealth. While he was still enraptured by her he wrote glowing letters to his daughter Sally enumerating these similarities. But his opinion quickly changed after he married her. It soon became apparent that Madame la Comtesse de Rumford-Lavoisier, as she preferred to be called, was an avid socialite who adored society and sumptuous dinner parties. Her numerous guests would constantly crowd the drive of their estate with handsomely appointed carriages.

Rumford, on the other hand, preferred his study and his laboratory, where he performed his beloved experiments. "I have had the misfortune to be married to the most imperious, tyrannical, insensitive woman who ever lived," he finally wrote to his daughter. In the end, they separated. He left her magnificent property in the Tuilleries and moved to an equally beautiful villa with a high-walled garden in Auteuil near Paris. He died there in 1814.

Rumford's Findings

Before I proceed with the details and events surrounding Rumford's discovery of the first main law, let me briefly touch upon the following details which appear to be of special importance in our investigation.

I have mentioned that Rumford achieved much fame as a benefactor, not only in Bavaria but also throughout Europe. He can be considered an outstanding example of a new breed of benefactor emerging at the time: a public-minded individual who put rational and scientific principles to work in the solution of social problems. He put his extraordinary inventiveness and scientific curiosity at the service of this very modern frame of mind. Rumford made significant improvements in the design of the stove. He invented a more effective reading lamp. He devised the first modern guidelines for a more cost-effective and nutritious diet, and was one of the first proponents of the soup kitchen. In Munich he was able to prepare nutritious and very inexpensive meals for 1,200 people in one sitting. In 1797 he served another meal at a mass gathering in Berlin at which 1,800 people were fed. His Rumford soup and his Rumford bread became known all over Europe. He devised new recipes and methods for producing wholesome and cost-effective bread. He actively promoted the consumption of the potato, which was still relatively unknown at the time, and contributed to making it a household staple.

During a short visit to Ireland he immediately began to draft elaborate plans to reduce the country's poverty and to help combat its cultural backwardness. In short, he was one of the first truly modern social reformers. His preoccupation with technical innovation and rationally guided social reform was constant and varied.

Rumford died at the age of 61. The great zoologist Georges Cuvier delivered the eulogy.

Rumford's Discovery of the First Main Law

We will now turn our attention to his discovery of the first main law of thermodynamics, which encompasses the idea that mechanical force, heat, electricity and chemical processes can be understood as different manifestations of the same indestructible, interchangeable energy. A report of his discovery was presented to the Royal Society in London on January 25, 1798.

As minister of war, Rumford exercised absolute control over the artillery workshop where the Bavarian military manufactured its arsenal. It was at this factory that he supervised the forging and drilling of cannons. In order to bore a shaft in the barrel of a cannon, a long drill was used that was powered by horses. Heat was produced at the point of friction where the drill bit bore into the metal. When Rumford put his hand to the barrel he was surprised by this heat and sought to understand its relationship to the work performed by the horses. He then proceeded to measure the heat produced in the metal. Rumford built an open wooden box, which he sealed onto the barrel's surface and filled with water. Lo and behold: after two and a half hours of drilling the water began to boil.

After careful observation and making a series of deductions, Rumford concluded that the heat could only have been generated by friction, which meant that it was caused by movement. Consequently, there had to be a physical connection between movement and heat. In principle this conclusion contained the essence of the discovery that later became known as the first law of thermodynamics.

Rumford then wanted to know just how much heat was generated by the drilling operation. Better still, he wanted to specify the exact quantitative relationship between heat and power. In order to answer this question, he lit the exact number of wax candles needed to boil the same amount of water that had been heated by the drilling operation. He found that "nine wax candles, each three quarters of an inch in diameter and burning together with a clear bright flame" produced the same amount of heat as the drill boring the barrel shaft of the cannon. Since the bore was powered by one horse, he was able to conclude that the power of one horse, or one "horsepower," was equivalent to the heat produced by the burning of nine wax candles. No one before Rumford had ever thought of directly comparing such disparate sources of energy.

Shortly after reaching this finding, Rumford found a way to express the relationship between heat and energy numerically. Later calculations proved that his equation was more or less correct. The article that Rumford presented to the Royal Society on January 25, 1798, is a most interesting historical document that offers us much food for thought.

First Annotation to Rumford's Article

In retrospect it seems odd that Rumford was so surprised by the fact that the drilling operation heated the barrel of the cannon. The fact that friction produces heat, after all, is a given of daily life. What child has not blistered his hand sliding down a pole or by letting twine slip too fast past his fingers while flying a kite? As an 18 year old, Rumford had sent his kite into an electrical storm and very nearly injured himself. Thus he had more than a slight acquaintance with kites and was probably also familiar with other sources of heat produced by friction. And had he never heard stories about people producing fire by rubbing sticks or stones together in the presence of dry grass? Gaston Bachelard, in his book *The Psychoanalysis of Fire*, gives numerous examples that show the universality of a practice that cannot have escaped Rumford's attention. It is no exaggeration to call the discovery of the relationship between heat and friction one of the first and most basic of all human discoveries. It was what made human civilization possible in the first place.

How, then, do we explain Rumford's surprise when he observed this known phenomenon in a new setting? As he wrote in his report: "I was struck by the very considerable degree of heat that the brass gun acquired in the short time of its being bored." It was this surprise that gave impetus to his discovery. Yet he also noted the fact that he had not been surprised. He wrote: "there was in fact nothing that could justly be considered as surprising in this event." In trying to reconcile these two observations Rumford then proceeded to qualify the first statement and reinterpreted his initial surprise as no more than "a childish pleasure" that was perhaps not entirely appropriate to the situation. But he then notes that when he called the workers' attention to the phenomena, they also were highly surprised, just as he was. It appears, then, that the production of heat by friction observed in the artillery workshop stood out in some manner as a new phenomenon. It was a phenomenon that differed from what happens when a boy slides down a banister or an old man rubs his hands together against the cold on a winter day. Such primordial experiences and reactions occur quite naturally and require no explanation. But what took place in the arsenal in Munich was a new event, one that drew Rumford's attention and set him on a new course of discovery.

To understand Rumford's initial surprise we should take into account the entire situation in which it occurred and not limit ourselves to the particulars surrounding the experiment. It is obvious that we should first pay attention to the workshop, the cannon and the drill. But we should also ask ourselves whether Rumford's surprise and that of the workers was perhaps generated by a change that took place in the metal itself. Could it be that metal underwent a fundamental change in Rumford's time? And did this change perhaps affect other materials in a similar way around the same time? These are certainly strange but legitimate questions, and they require a thoughtful response.

As such they lead us to ask another more straight-forward question. What other changes occurred in Rumford's time when he made his famous discovery? What events were synchronous with his scientific breakthrough?

Second Annotation to Rumford's Article

Rumford performed his experiments in the years 1796 and 1797. Historians of science have generally recognized that researchers plan and execute their research only after they are at least partially convinced of what the outcome will be.

That means that at the end of 1795 or at the beginning of 1796, Rumford must have known that he was confronting a changed world and that he was dealing with an altered sort of metal. He may have harbored these suspicions without necessarily thinking consciously in such terms. Nevertheless, Rumford obviously intuited an important shift in the material world and this allowed him to hypothesize about heat, energy and power in unprecedented ways. The question remains then: What else happened at the end of 1795 or at the beginning of 1796 that could have triggered Rumford's intuition? It is not always easy to determine what is contemporaneous with an event because a historian does not necessarily work within the same paradigm as the metableticist. In this case, however, the answer is obvious.

In 1795, the French Revolution was still underway. It was followed with intense interest by the rest of Europe because the French Revolution was essentially the revolution of the whole of the Western world. The initial fervor of the Revolution had abated, but it continued on its course. Jean-Paul Marat, who had been buried in the Pantheon, was removed from his tomb. Napoleon Bonaparte was already waiting in the wings. French troops had landed in foreign countries, where the ideology of the French Revolution spread in spite of all opposition to it.

What was the ideology of the French Revolution? It may be summarized in three words, which resounded and echoed far and wide: Freedom, equality and fraternity. From our present perspective over two centuries later, it becomes clear that of these three words, the word *equality* carried the most ideological weight.

So the question we need to ask is this: Did this political battle cry of equality have anything to do with Rumford's strangely transformed metal? Antoine Lavoisier, whose widow Rumford had so disastrously married, was also an intellectual contemporary of Rumford. In 1789, the year of the storming of the Bastille, Lavoisier published a book on the chemical elements (*Traité élementaire de chimie*), in which the classic elements of fire, air, water and earth were summarily dismissed. These were replaced with multiple elements, which banished any mystical, magical or anthropomorphic characteristics in which the ancient elements had been steeped. The ancient elements, water, air, fire and earth, were *local*. Water from the Seine was not the same as water from the Danube. Not because they contained different levels of pollution or were differently constituted, but because they were different *as such*.

This of course hasn't changed. You would only need to swim in each river to know that they are different. But as a result of the vision introduced by Lavoisier, these differences are no longer considered essential. The new vision of the elements does not leave any room for local differences. There would be differences that for some time would remain enshrined in folklore and poetry, but the new elemental viewpoint would show these differences to be cosmetic. Within this new world revealed by Lavoisier iron would be iron, no matter where it had been mined or refined. Chemical additives certainly would distinguish one batch of iron from another, but iron itself would cease to be a local product in the old sense.

Rumford's enormous public meals were prepared with this newly "equalized," or generic, water and other "equalized" ingredients. These mass events offered hundreds of participants "equal" meals of "equal" soup and "equal" bread. Meals were the same whether they were served in Berlin or Paris, with every participant eating the exact same “Soupe à la Rumford” and “Pain à la Rumford.”

“This doesn’t taste very good,” people would say to each other when sitting down to one of these meals. Equalized soup and bread tasted rather horrible, but they were very well constituted ideologically. Do our contemporaries realize they are betraying the revolutionary cause of equality in the simple act of buying a loaf of fresh, locally baked bread?

The Abolition of the Guilds

At the beginning of the French Revolution, all guilds were closed and prohibited. This was done in the name of “freedom of labor.” This regulation, which ended the age-old tradition of guilds, was called the “Proclamation de la liberté du travaille.” What were guilds?

Guilds existed in all countries. Guilds were prevalent in China, India, ancient Rome and also in Europe since early medieval times. Tradespeople and merchants formed occupational associations to foster professional fellow feeling and brotherhood. Fellowships of butchers, saddle makers, smiths, fishmongers, and most certainly bread bakers. They were formed to protect workers against unfair competition and price gouging while maintaining a sound standard of quality. Every guild had its own religious festivals. The protective role of the guild was generally positive. Jealousy and competition were quelled to such an extent that associates of a guild lived on the same street, the names of which are a litany of the old guild trades: Saddle Street, Smith Street, Baker Street.

Now the bread from a baker on Baker Street of days past was bread that was intrinsically and literally different from Rumford’s bread. Not because the guild’s bread could be called specialty bread and Rumford’s bread could be identified as soup kitchen bread, but because their social structure, the ideological and cultural underpinnings of each loaf, was entirely different.

The Mystery of Commodities

Whoever doubts this difference — and it is very easily doubted! — may find it advisable to consult the book in which Karl Marx (in 1897) not only propounds the basic tenets of historical materialism but also unveils something that could be called the mystery of commodities. The mysteriousness inherent in every commodity can be found in the two aspects of any commodity that can be bought and merchandised. Let us take, for instance, bread, an example deliberately chosen by Marx. As long as bread remains in the possession of the baker, it is "incorporated" labor. This refers to labor that emerges out of human circumstances and relationships. It emerges out of a relationship between a master and servant, or a baker and his assistant. This labor not only actualizes the relationships between people but also makes time concretely visible in, for instance, the time needed to form ingredients into bread, to mix water with flour, to knead the dough and to form it into loaves. It also includes the time required to insert the loaves into the oven, supervise the baking, and take the bread out of the oven. Thus bread is an artifact of *festgeronnene Arbeitszeit*, or "frozen labor time," according to Marx's apt formula.

Even more apt is his definition of bread as "a relationship hidden under a material cover": In short, a relationship between master and servant, or persons performing labor. But the same bread, baked and finally placed on the shelves of the grocer, is subject to a "metaphysical" change as soon as it is bought. Merchandise is suddenly converted into consumer goods, with a new value determined by the market of supply and demand, thus losing its capacity to be the carrier of human labor relationships. Thus, bread is similar to every other commodity — a "very distorted thing, full of metaphysical subtleties and theological conceits" — a magnificent definition that likewise stems from Marx.

New Bread and New Metal

The French Revolution ushered in pernicious, metaphysical, even theological changes. Rumford fed his new bread to the masses. The huge number of his guests (and the bread's awful taste) can be taken as irrefutable proof that his bread was perfectly modern. It was new.

Now for the leap which is really no leap at all. With his new metal — new iron, new bronze — Rumford discovered the law of the equality of heat and power. This law was discovered in the process of boring a cannon that — as we all know now — would subsequently feed on the lives of thousands, even millions, of people since the French Revolution and because of the French Revolution. No one had ever dreamed of Rumford's law before, all for the weighty reason that the preconditions conducive to such a law did not exist.

The newly transfigured metal that Rumford subjected to his investigation emerged from the old as the result of, to borrow Marx's terms, a metaphysical conceit or theological sleight-of-hand.

A theological sleight-of-hand? A metaphysical conceit? What then, exactly, was the nature of the uncanny theological alteration that had occurred in Rumford's metal? And in what way did this mysterious change set the stage for Rumford's experiment at the artillery? To answer that question we need to draw our attention once more to the French Revolution.

The French Revolution

As we mentioned before, the Revolution advanced under the banner of a newly declared equality. This ideology in action effectively put an end to the class state, even though many relics of it have survived into present times. One should remember that the class state functioned for centuries in Europe as well as elsewhere as a fully legitimate and just form of government. It then becomes clear that the French Revolution placed before itself a nearly impossible task that could be accomplished only with a great deal of violence. This violence was instituted in order to undertake the literal erasure of two classes: the priesthood and nobility, the aim being that only one class, the "third," would then remain. This reached its height in the Terror, which consisted of nothing more than the random killing of people who belonged to the classes slated for elimination, or anyone who made common cause with them. The Terror lasted from the summer of 1793 until the death of Robespierre in July 1794. It was in the following year, by the way, that Rumford carried out his experiments at the artillery.

By September 1792 the Republican calendar had already replaced the Christian one. In order to entirely de-Christianize the state, or to rid it of its divine Head since the Head (note the capital *H*) had certainly never belonged to the third class, the religion of reason was put in its place. For Robespierre, the leader of the Terror, this took things too far. He did not think this was so for religious reasons. He merely feared that the Revolution might degenerate into chaos. To prevent this he wanted to institute a religion of reason and declared his allegiance to a Supreme Being.

On June 8, Robespierre, his face strained with emotion and suffering more than ever from his habitual nervous tic, appeared on a podium in order to announce the change in a tremulous voice. On June 8, the Festival of the Supreme Being (fête de l'Être suprême) took place: in which the name of this remote Deist figurehead was nonetheless still capitalized. In the garden of the Tuilleries, a great platform was erected — an old drawing of this is printed here. Seats for the government were located behind the podium. Flags embellished with medallions, embossed with the letters RF (République Française), hung on the pillars. A square banner loudly displayed the maxim "Liberté ou la mort" (freedom or death) — the slogan of the Terror.

Platform in the garden of the Tuilleries that was erected for the Festival of the Supreme Being.

Robespierre arrived on the platform late. In the meantime his cohorts observed the waiting empty chair and mumbled audibly to each other, “Il fait le Roi!” (He is playing the king!), the ultimate insult at that time. Despite these murmurings, a festive air of pomp and celebration prevailed. The people pouring in for the event looked absolutely stunning: the ladies were most elegantly coifed and appareled. When Robespierre eventually made his appearance, he was resoundingly applauded. After delivering a short speech, the huge crowd proceeded to the Champ de Mars, which can still be seen today between the Eiffel Tower and the Military Academy. The freshly commissioned hymn addressed to l’Être suprême was then sung.

Robespierre apparently felt wretched after the spectacle. Despite the high pomp and acclaim of the event, it was fairly obvious to bystanders that his strategy had backfired. Softly but audibly his associates could be overheard hissing, "Look at the man, he wants to be God him-self," "High priest," "Brutus still exists." When Robespierre arrived home, he told a confidante, "I will not be seen among you much longer."

The Field of War, or Champ de Mars, became the festal field of the French Revolution. It is not generally known that in September 1798 (the year when Rumford's article on the first law was made public), the first ever French national industrial exhibit took place. It vaunted the fruits produced through the "liberation of labor" proclaimed at the Revolution's beginning in 1789. "Liberation of labor" meant liberation from the shackles of the guilds, of course. With the completion of this course of events the way for Rumford's discovery had been cleared.

Marx's "Theological Sleight-of-Hand"

But where is the theological sleight-of-hand in all this? It can be seen in the old drawing of the amphitheater where Robespierre spoke. Whatever his opinion about the excesses of the religion of reason, his cult of the Supreme Being was meant to serve as a ritual exorcism of the unbridled murder and manslaughter of the Terror. This was nothing less than blasphemy. It was Marx's "theological sleight-of-hand" in action. The inhuman agenda of the Terror was concealed behind a pseudoreligious mask.

With that "theological trick," metal changed in ways that made the discovery of the first main law of thermodynamics possible. On July 28, 1794, just seven weeks after the Festival of the Supreme Being, Robespierre was beheaded.

A Rolling Mill

Years ago I was afforded the opportunity to visit a steel mill and see how steel is forged and tempered into rails. Once there

I found myself on an elevated, Plexiglas-enclosed platform, which overlooked an immense and lengthy hall with what seemed an endless series of hot rollers and conveyance rollers running its length.

Emerging from a nearby forge that was concealed from my vantage point, a great mass of glowing steel in the shape of an elongated cube was suddenly thrust onto the conveyance rollers. A rail had to be tempered from this block of steel. The steel was then conveyed onto the hot rollers where it was compressed and elongated into 50-meter sections of train and subway rail.

The process was drowned in hellish noise. The steel was continuously conveyed and tempered. Any human participation in the process was invisible; nevertheless, the process was being overseen and guided by persons I could not see. During the entire process, the steel remained a glowing white. That is, while the steel continued to release heat into the atmosphere, rolling onto the conveyance rollers and then onto the hot rollers, the temperature of the steel remained nearly constant. This made quite an impression on me but nevertheless elicited no surprise. During my secondary education, I had to learn about the law of conservation of energy, which obviously implied that the quantity of energy needed for the tempering of the steel released a large amount of energy in the form of heat. There was no reason for me to be surprised — although one did have to know the first law of thermodynamics and preferably should not know too much about the initial surprise of which Count Rumford wrote.

Nonetheless, what happened during the tempering was very impressive, and remains vividly so in my memory. I can still see the glowing white of the steel pipes. I remember the pain I felt in my eyes as I looked on. Add to these impressions my childhood memories of the pain I felt on my palms when I slid down my aunt's banister too quickly. The memory of the pain in playing out a kite string too quickly through my fingers. The mystery associated with these experiences still remains. I did not become estranged from such mysteries when I became acquainted with the first main law of thermodynamics as a young student in high school. Now I know for certain that knowledge of this law can never completely obscure these mysteries, unless one unconditionally surrenders to the premises of natural science. Now I know that the mysteries remain, and that natural science is not the science of the whole of nature.

A Question about Natural Science

In my younger years I became a passionate phenomenologist. To approach the world as a phenomenologist means to resist our reflexive impulse to seek out the cause that gave rise to an event and to give ourselves the time to fully describe the phenomenon as it unfolds before us instead. It is a way of approaching the world that asks how? instead of why? During my training as a medical doctor, I acquired the almost unbreakable habit of confronting any unusual situation with persistent questions about its underlying or preceding causes. Phenomenology presented me with another way of thinking and of asking questions. In psychiatry, this approach proved very useful. In talking with psychotic patients we are little helped by posing medical or biological questions about the causes of their hallucinations. Such a natural scientific frame of mind does not predispose us to better understand or to better relate to the patient in front of us.

To ask a very different question, to inquire into *how* it is that a patient hallucinates and to seek to know in what manner this differs from ordinary ways of seeing, hearing and understanding presents us with an equally difficult and time-consuming task. But this line of questioning has the advantage of bringing us closer to and in more immediate contact with the patient.

The work of the phenomenologist has often been compared

to that of a painter who also goes through endless trouble to describe or depict the "how" of a particular world. We may think, for example, of Cézanne's tireless effort to depict Mont Sainte-Victoire in all seasons and at all hours of the day. The magnitude of this task would at times overwhelm the painter and make him behave in strange ways. It is said that at one time the painter was disturbed in the middle of his work by the crowing of a rooster. Cézanne put his brush down in disgust, threw the unfinished canvas onto a nearby hedge and stomped home.

To understand his irritation we must assume that the rooster's crow had somehow changed the appearance of the mountain: that it had changed the manner it unfolded before the eyes and ears of the painter. Cézanne did not want to paint the mountain's new manner of appearance. He got up and went home. This gesture may strike us as a bit extravagant but seen within its context it remains quite understandable, nevertheless.

We all may agree that the crow of the rooster had changed the appearance of Mont Sainte-Victoire, and that it had modified it in some manner so that the painter was confronted with a slightly different landscape. Cézanne had painted it in all seasons and under all circumstances. He had studied it in autumn and spring, had seen it under a flood of sunlight and obscured by thunderclouds, had contemplated it with or without the crow of a rooster. But what about Mont Sainte-Victoire itself? What is the "it" that changed its appearance under all these circumstances and yet remains the same?

A natural scientific answer to that question would maintain that the mountain consists of stone, soil, leaf mulch, botanical matter, and so on. We might add still further details to this naturalistic description. We might say that the mountain consists of a particular kind of stone, of bauxite, and that this compound is found throughout Provence. We might further specify that the chemical formula for bauxite is Al_3O_3 and that the aluminum atom can be further thought of as a solar system on a very small scale. This system contains a minuscule sun called a proton (in the case of aluminum, a clustering of little suns) surrounded by a quantity of planets called electrons. Or we might think of that atom purely in mathematical terms and in that case Mont Sainte-Victoire would ultimately be best represented in the form of pages filled with mathematical formulas.

The metableticist does not dispute these mathematical, chemical or geological interpretations of Mont Sainte-Victoire. He or she would point out, however, that bauxite is not the ultimate essence of the mountain, but merely a precondition for its appearance. Bauxite is to the mountain what H_2O is to the river, a precondition for the appearance of Mont Sainte-Victoire, the Seine and the Danube. Moreover, one bauxite mountain range differs from another in the same way that water in the Seine differs from that in the Danube. We are really not much closer to answering the question as to what Mont Sainte-Victoire is in its final instance.

This does not imply a criticism of natural science or a failure to admire and honor its achievements. But neither does it imply closing one's eyes to the limits and even the dangers that are inherent in this way of thinking.

The development of natural science was the work of countless individuals who doggedly strove to solve recalcitrant problems. It has often been the work of singular persons. There is a good fit between singular individuals and recalcitrant problems.

We mentioned Count Rumford as a singular personality at the beginning of this chapter. The problem posed by heat generated in drilled metal was itself a strikingly singular one. The first main law emerged from the encounter between a singular person and a singular phenomenon. However, Rumford was not the only one to discover the first law. That law was discovered again at a time when it had been almost forgotten. The next chapter will deal with another singular person who discovered the first law.

CHAPTER II

The First Main Law: Julius Robert Mayer

Count Rumford's work on the first law of thermodynamics gradually fell into oblivion. Until recently there appeared otherwise accurate accounts of the history of thermodynamics that did not mention his name or make reference to his discovery. This strange omission and disregard of such an important scientific breakthrough suggests that Rumford's vision did not quite fit within the existing conceptual framework of the time. The equivalency of heat and energy suggested by his experiment must have appeared strange and unacceptable to his contemporaries. Perhaps this was particularly so because his discovery relied so completely on the direct and unaided observation of a common, everyday phenomenon. It made no reference to hidden, invisible or underlying forces, nor did it use mathematical reasoning to reach its conclusions. In fact, it was only in the middle of the nineteenth century that a connection was made between the first two laws and molecular motion. It is in any case noteworthy that the first law was discovered by a scientist who relied so completely on the direct and unmediated observation of an everyday common event.

Julius Mayer was also a scientist whose work was inspired by direct observation of everyday events, and his work also remained unacceptable to his colleagues for a long time. He was not relegated to obscurity, as was Rumford, but it was his fate to be almost universally ridiculed and belittled.

We do not know whether this lack of regard was the source of the despair that drove him to throw himself out of a second story window and seriously injure himself. What is important to keep in mind, however, is that the first law of thermodynamics was once again discovered by a scientist with a great flair for concrete and visual observation of daily events.

Julius Robert Mayer was born in 1814, the son of an apothecary in Heidelberg. In 1838 he received his medical degree and began practicing family medicine in his hometown. He did not pursue private practice for any length of time. In the winter of 1840 Mayer became engaged as a medical officer for the Dutch East India Company and boarded a three-mast schooner, the *Java*, en route to the Dutch East Indies. The ship arrived in Batavia after sailing for 101 days. From there, it went on to Surabaya, and then sailed back to the Netherlands with a load of sugar and coffee.

Mayer kept a journal during the voyage that brought him from Rotterdam to the Indies, but he did not continue the practice on the way back home. He wrote to his parents that on that particular journey he had been preoccupied with study and meditation instead. A couple of years later, he also wrote the psychiatrist Wilhelm Griesingen, a friend from his student days, that during that particular time he felt uniquely inspired, in a way that was never to be repeated in his life again. This inspiration had been sparked by something he had observed shortly after his arrival in Batavia. It was this observation that started him on a long train of thought that led to his discovery of the first law.

Mayer's Experience in Batavia

Like any physician of his time Mayer frequently performed bloodletting on his patients. It was while performing this standard procedure that he noted something unusual — namely, that the blood drawn from the veins of his European patients in Batavia was nearly the same color as that drawn from their arteries. This was odd, because in Europe it was commonly noted that arterial blood was darker in color than venal blood. How could one account for this difference? Mayer reasoned that in the colder climate of Europe there was a more marked difference between the temperature of the environment and that of the body. To maintain normal body temperature in such a climate it is necessary to insulate the body and protect it from the cold. Mayer also remembered from his reading of Lavoisier that the body maintains its own normal temperature by means of an internal combustion process. Lavoisier had suggested that this internal combustion process required oxygen.

Oxygen is transported by blood through the arteries from the lungs to the various organs of the body, where it fuels the combustion process. The veins then transport the oxygen-depleted blood back to the heart. Blood that is rich in oxygen is lighter in color than blood that is oxygen-depleted. That is why arterial blood is much brighter in color than venal blood in a colder climate. But in the Tropics less combustion is needed to maintain body temperature, and blood returns to the heart still rich in oxygen. That is to say, it remains brighter red in color. This phenomenon becomes even more pronounced if the body has not performed any physical labor beforehand.

It was this phenomenon that prompted Mayer to think about the relationship between heat and work, and finally led him to discover that both phenomena could be considered manifestations of the *same energy*. This was the exact same conclusion reached by Rumford when he observed the drilling of cannons in the Bavarian artillery. It appears that Mayer was able to reformulate the first law of thermodynamics without having had any knowledge of the work of its first discoverer. All that remained for him to do was to describe the numerical relationship between heat and energy. Mayer succeeded in doing this as well.

Mayer was not alone in observing the difference in color between the blood of veins and arteries, and the lack of such a difference in the Tropics. Most every doctor of his time had knowledge of this fact. The originality of Mayer's mind did not lie in the fact that he noted the difference but in the fact that it set in motion a train of thought that revealed the world in a new light. His train of thought was in fact so original that even today, some 150 years later, we still have difficulty following it.

"The Ocean Warmed and Flayed by Storms"

Spurred by his original insight, Mayer started the 121-day voyage back to the Netherlands. He no longer thought back to the many impressions he had received during his stay in the Tropics. He no longer kept a diary. Mayer used these days at sea to think exclusively about the possible connection between heat and power, which he thought was impressively demonstrated by the ocean surrounding his ship.

It had been known for centuries that fierce winds and storms cause the temperature of waves to rise. No less of a figure than the Roman orator and writer Marcus Tullius Cicero noted this very fact in his remarkable book, *On the Nature of the Gods*. Cicero also provides an explanation of sorts for this warming effect: "The warmth of the waves should not be understood as coming from the outside but it is churned up by the storm from the deepest depths of the ocean where it is always warmer."

Mayer had never heard about warmth being generated by the wind. But one of the crew on board did know about it, and related the phenomenon to Mayer. This set him thinking, until he came to the conclusion that it was not the hidden warmth of the ocean that warmed the waves, but the wind itself, which released its energy to the waves and manifested itself as heat.

The Flask of Shaken Water

Mayer must have had considerable difficulty in accepting his own hypothesis. This is confirmed, in fact, by an incident that occurred at a lecture Mayer delivered several years later. While he was propounding his insight that movement and heat were "one and the same object in different forms of appearance," a member of the audience skeptically remarked: "Then water should become warmer after being shaken." Mayer left the hall without saying a word — only to turn up a few days later at the heckler's house and declare, "It is so." One can only conclude that this errant remark renewed his ancient doubt and drove him to conduct more experiments at home in order to verify his earlier findings once again.

Mayer summed up his discovery in an article he sent to the prominent physicist, Johann Poggendorff, requesting that it be published in the *Annalen der Physik und Chemie*. Poggendorff sent no reply and even neglected to return the article upon the repeated requests of its author. Mayer finally sent another copy to Liebig's *Annalen der Chemie und Pharmacie*, where it was published in 1842 under the title *Bemerkungen ueber die Kräfte der unbelebten Natur*.

This was nearly half a century after the discovery of the same law by Rumford. At approximately the same time, the law was also discovered and published by Colding (1843), Joule (1843) and Hotzman (1845). This is not so surprising by itself. What is striking is that Rumford should have been so alone in his discovery and that it had so little impact. One can only conclude that the political backlash following the French Revolution created a climate of thought that was adverse to any further exploration of Rumford's insights. This situation changed around 1840, when his train of thought was taken up and pursued once more.

Both Rumford and Mayer made the *same* discovery. Rumford's discovery, however, was synchronous with the French Revolution, after which it was forgotten for almost 50 years. What events surrounded the law's rediscovery? To find the events that were synchronous with Mayer's discovery, we must ask, "What exactly was happening in the social political arena of Europe in Mayer's time?"

Proudhon and Marx

In 1840 Pierre Josef Proudhon published a book entitled *What Is Property?* (Qu'est-ce que la propriété?) Proudhon's answer was: "Property is theft." Proudhon wrote: "When someone asks me, 'What is slavery?': I would answer in a single word, 'murder'! And everyone should understand why. And if someone should ask me 'what is property,' I would also respond with a single word: 'Theft.'"

In a slender volume Proudhon sets forth a new differentiation between ownership and mere possession. According to Proudhon, possession pertains to things we use and relationships we have in our immediate sphere of activity. We possess a knife and a fork, or a table and chair. In Proudhon's time, this notion of possession also extended to one's children and wife. Thus possession refers to all objects and relationships we come to "possess" through the exertion of effort or labor of one sort or another. If this is possession, however, what is *ownership*, and how does it differ from *possession?*

Proudhon defines *ownership* as the holding of land, real estate and the means of production. This refers especially to property and land that is owned without effort or direct labor: For instance, it refers to any land, resource or means of production that is passed through inheritance from father to son. Proudhon's book met an understandable wave of criticism and discontent, but he was spared any persecution or legal prosecution because he enjoyed protection at the highest levels of power.

This book is the cornerstone of all subsequent leftist ideology. It is the foundation for all leftist rhetoric, leftist theory, and it set the stage for such revolutions as the October revolution of 1917. None other than Karl Marx wrote a favorable review of Proudhon's book in the *Neue Rheinische Zeitung* in 1842, thus introducing and popularizing Proudhoun's ideas for the first time outside of his native France.

But Marx himself took direct part in what could also be termed an event synchronous or simultaneous with the rediscovery of the second law. As the editor of the *Rheinische Zeitung*, he was struck by a growing crisis: the exploding number of wood thefts that were occurring between the years 1842 and 1843. Judicial records show that in Prussia alone, five out of every six court prosecution cases were for theft of wood. This was wood allegedly stolen by common folk from forests that were owned either by the crown or by nobles.

What was odd was that this "crime" went virtually unreported prior to this time. This was due to the fact that the gathering and chopping of wood by common folk was not generally considered to be a criminal offense before 1842. For centuries, in fact, it had been the unquestionable right of the common man to chop wood in a local forest of an owner. The title "owner" is even a bit of a misnomer, since the land-holders were also frequently perceived as the protectors of those dwelling on their estates.

But wood had become scarce and therefore highly valuable by Marx's time, encouraging "owners" to sell their wood. The former woodchoppers exercising an ancient right found themselves suddenly transformed into "thieves."

But what was responsible for the rising cost of wood in Germany and throughout Europe? It was brought about by the new demands of the Industrial Revolution, which required wood to be burned in the furnaces of its ever-growing number of steam machines. This demand changed wood that had been publicly shared according to ancient custom into privately owned raw material for sale on the industrial market. It was Marx's conviction, however, that the gathering of wood should never be referred to as "theft" because some things should never be allowed to become the private property of individuals.

Marx could have supported this argument with his later explanation of the difference between merchandise and things that are kept for personal use. I have already touched on his ideas concerning the various forms of commodities, ideas that could elucidate the commodification of chopped wood. Industry transmutes wood intended for personal use in fireplaces and stoves into industrial raw material for furnaces in factories. This change of wood from object of personal use into merchandise concerns a fundamental change in the nature of wood. A change that cannot be observed in the wood's chemical structure, yet one that has all-pervasive consequences.

It follows from this salient example that, at the time of Mayer, Proudhon and Marx metaphysical changes took place not merely in iron and bread but in innumerable substances and things. It was occurring in sand, lime, cement and, of course, in wood. An entire book could have been written on this Marxist maxim alone. Such a book would deal directly with fundamental changes that take place within the very substance of things. It would describe the phenomenon of transubstantiation.

Do we have to assume that Mayer, pensively observing the storm-flayed ocean from the deck of his ship, was beholding "metaphysically transformed" seawater? Most certainly! Does it not also follow that Mayer was observing metaphysically transformed blood when he cut into the veins of the patients he treated on the journey from Europe to the Dutch East Indies? No less certainly. Let us note that the same subtle metaphysical transformation of blood had first been observed by such scientific geniuses as Vesalius and Harvey. Genius is necessary in order to perceive the metaphysical changes in things. Rumford was such a genius, and so was Mayer.

The word *genius*, however, is much too broadly defined and vague to describe the uniqueness of such people like Rumford and Mayer. Earlier I noted that they were "visionaries." They do not merely look at the world; they transform the world as they see it. They imbue what is seen with thought. While observing they design presuppositions and structures that underlie what they see. They imbue the seen with an initially dubious hypothesis, which gradually solidifies into a definitive theory.

At the risk of being misunderstood, it must be said that visionaries thrust and implicate themselves into the thing that they observe. That which they observe invites and beckons them to the point where they are compelled to put all of their surprise and expectation into it. Their imagination is aroused at the object's invitation, by what is seen. It pulls the whole of the visionaries' waking dream, his ardor, into itself.

Every discovery comes into being through the meeting of two sides. It arises from the domain of a mysterious thing that beckons from afar, and is answered by the visionary who goes forth to meet it.

CHAPTER III

The Second Main Law: Carnot, Clausius, Kelvin

Rumford and Mayer, the discoverers of the first main law, were seers or visionaries. They relied on their powers of natural observation to discern a shift in the nature of things. The discoverers of the second main law, however, relied less on this type of intuitive and insightful observation. They were more inclined to mathematical and rigorous scientific abstraction, working out the quantitative and mechanical details of natural processes on paper and in the laboratory. This variance in scientific approach reflects the different natures of the first and the second law. As a general rule it is possible to draw a parallel between a particular type of scientific problem and the type of scientist it attracts to explore and eventually resolve it.

In the following chapter it will also become clear that the two laws represent different types of scientific problems. It also will show that the historical synchronisms surrounding the discovery of the first law of thermodynamics differ from those that surrounded the second law.

History records three scientists as chiefly responsible for the discovery of the second law: Carnot (1824), Clausius (1850) and Kelvin (1852). Of these three, Carnot occupies a special place. Although he did not actually discover the second law, he nevertheless should be seen as responsible for the preliminary groundwork that led to its discovery. To understand the discovery of the second law in its proper historical context we need to draw upon a brief biographical sketch of each of these three individuals, together with a summary of their respective contributions toward the discovery of the second law.

There are a few essential concepts mentioned previously that we should keep in mind before we proceed. The second law states that in the transmission from heat to mechanical energy, temperature always decreases and that a substantial amount of this heat is not transformed into mechanical energy, but is "lost" due to scattering and dissipation.

Sadi Carnot

What strikes us first of all about Carnot is his unusual first name. The father named his son Sadi after the thirteenth century Persian poet whom he deeply admired. Sadi was born in 1796 in a wing of the Palais du Luxembourg in Paris. His family had taken up residence there since his father, an engineer and a mathematician, occupied an important position in the second revolutionary French government.

Sadi Carnot studied at the Ecole Polytechnique and graduated at the age of 24. He subsequently undertook to visit all the major industrial centers of Europe in order to acquaint himself with technological practices current at the time. In his twenty-eighth year, in 1824, he published a slender volume of 118 pages that proved to be one of the most important and influential publications of the century. It was entitled *Réflexions sur la puissance motrice du feu et sur les machines propres a développer cette puissance*. Translated this roughly means: "Reflections on the driving force inherent in fire and on the machines capable of using this power."

Why did the book attract such attention? What was the central idea that animated it? The central idea states quite simply that the capacity of fire to generate mechanical power depends on the transfer of heat from a warm body to a cold body. In other words, fire alone is not sufficient to power a machine. An essential component of the generation of mechanical power from heat is that of the transfer of heat from its source to a colder body. It is this essential component that constitutes the working principle of steam and gas powered machines. They all function by the grace of what we could call the Carnot principle.

This understanding did not follow immediately upon the publication of Carnot's book, which had a curious history. Very few scientists appear to have been aware of its publication. In fact, it appears to have been forgotten to such an extent that the physicist Kelvin, who knew of the book only from hearsay, unsuccessfully attempted to get hold of a copy in Paris in the early 1840s. Even some 30 years after its publication, and after the book had been republished in limited editions and translated into several languages, it was still not widely read and remained difficult to find.

Direction

However important Carnot's principle may be by itself, still of greater importance is his discovery of the *direction* of a natural process. For the first time in the history of natural science, Carnot's principle makes explicit reference to this *direction* as itself a necessary and essential part of a natural phenomenon. For heat energy to transform into mechanical energy it must travel a one-way path from hot to cold. This path and this polarity should not escape our notice.

It is unfortunate that little more can be related about Sadi Carnot, however. He died of cholera at the age of 36. Hippolyte Carnot donated the entire body of his brother Sadi's work to the Academie des Sciences in 1878, along with his unpublished notes. These notes belatedly revealed that Sadi should have been credited with the discovery of the same law that Clausius had rediscovered in 1850.

Clausius

Rodolphe Clausius was born in 1822 in Koelin, northeast of what was then known as the Prussian city of Stettin, now a part of Poland. He was the youngest of 18 children. His father, an inspector of primary education, tutored him up to the highest two classes of the gymnasium, which he then completed at Stettin. He studied physics in three consecutive cities: first in Berlin, then in Zurich, then in Bonn. He was married soon after, and according to his biographer, F. Folie, his household became a haven of happiness and "Gemütlichkeit" for his family and friends. He was fond of playing the piano and of singing with his wife and children. One of their favorite composers was Kirchoff, the physicist who is known for his groundbreaking work on spectral lines. Clausius was diagnosed in 1888 to be suffering from pernicious anemia, an incurable illness at that time. He died that same year.

Clausius appears to have lived a completely ordinary bourgeois existence. His life is totally unlike that of the restless and adventurous Count Rumford. Nor was there anything in Clausius's temperament to match the restless excitement of Mayer, who was so mesmerized at the spectacle of storm-driven waves on his journey home from the Dutch East Indies.

The Significance of Clausius

In 1830 Clausius published an article in the *Annalen der Physik und Chemie* that, in our present context, certainly deserves our attention. It bore the long title: "Ueber die bewegende Kraft der Warme und die Gesetze, welche daraus für die Wärmelehre selbst ableiten lassen": (On the power of heat to generate motion and on the laws of thermodynamics that can be derived therefrom.) In this article he confirms Carnot's findings, namely that heat invariably streams from a hot body to a cold body — but he rejects Carnot's supposition that no heat is lost in the process. He maintains that a great deal of heat is actually lost or scattered in the transfer from heat energy to mechanical energy. It is not "lost" in the strict sense, but scattered or dissipated.

We need hardly mention that Clausius published a great many more articles than the aforementioned one. He accordingly enjoyed wide recognition and esteem as a great scientist. For our present purposes, however, only the 1850 article is important.

William Thomson

William Thomson was of Scottish descent and was born in Belfast, Ireland, in 1824. His father, James, was a professor of mathematics at the University of Glasgow and soon became aware that his son possessed extraordinary talents in mathematics and physics. He resolved to teach him and became his mentor. When William was 17, his father deemed him sufficiently ready to study at the University of Cambridge, then as now one of the most important universities in England. His father very skillfully tutored his son with the conscious aim of preparing him to occupy the chair in mathematics at Glasgow. At the age of 21, William published two articles in the *Cambridge Mathematical Journal*. This was a spectacular achievement and was obviously most edifying to his attentive father.

William moved to Paris in 1845, where his father had arranged for him to have suitable living arrangements and an introduction to the most important scientific and mathematical minds in Paris. Having arrived there, William struck up friendships with the best-known mathematicians and physicists of his time. This was quite easy for him to do since he was quite amiable and extraordinarily intelligent.

When the chair of natural philosophy (mathematics) in Glasgow became vacant, William's father tirelessly campaigned for his appointment. He was then appointed to the chair at the youthful age of 22. William proved to be a brilliant university teacher and indefatigably produced highly original articles.

In 1849, when William was 25 years old, his father died of cholera. Shortly afterward, William made his first offer of marriage, which was refused. He pressed his suit once more shortly after, only to be refused by his beloved a second time. It turns out that the lady secretly hoped he would sue for her hand a third time, but he failed to do so. In 1853 he got married instead to his cousin, Margaret Crum. The two had known each other since childhood, a fact noted in his biography, *Lord Kelvin, the Dynamic Victorian*, by H. I. Charlen. During their honeymoon in Switzerland, the young lady fell ill and remained so until her death in 1869. No children were born.

One hardly needs to be a psychiatrist to suspect that William Thomson was no Don Juan. He could be characterized instead as the dutiful son — father's "little boy" — though it should be noted that modest patronage by a father on a son's behalf was more common in his day, and certainly considered less dangerous than it would be now.

Second Marriage, Nobility and Kelvin's Death

Thomson married for the second time, two days before his fiftieth birthday. He then had a new residence built and left his paternal estate. No children were born to this marriage, either. In 1892, when Thomson was 68 years old, he was raised to the peerage with the title of Lord Kelvin of the Largs, Kelvin being a small river near Glasgow, and Largs the small hollow where his house rested. In the extensive literature devoted to him since that time he has been exclusively known as Kelvin.

Kelvin had been a professor at Glasgow for more than 50 years when he decided to vacate his chair at the university in 1899. He was 75 years old by this time and no longer fully in tune with the times. He accepted very little of the new physics, which was already in full flower. He refused, for example, to believe that nuclear fission was possible. He also refused to accept that radium could give rise to helium. He dismissed new ideas concerning the disintegration of the radium atom as "wantonly nonsensical," according to H. J. Sharlin's biography, *Lord Kelvin* (1979).

Kelvin was deservedly honored as one of the greatest physicists of his time. He was the founder of the absolute temperature scale, expressed in Kelvins. He invented the gyrostatic compass, an extraordinarily sensitive electrometer and the receiving apparatus for the telephone. All that aside, however, it is not possible to call Kelvin a visionary, like Rumford or Mayer.

Lord Kelvin was entombed in Westminster Abbey, close to Isaac Newton and not far from Charles Darwin. He belongs among the great historical figures of Great Britain.

Kelvin's Contribution to Thermodynamics

Our discussion of Kelvin's contribution to thermodynamics will be limited to an article he wrote in 1852. In this article he arrives at the conclusion, based on Clausius's observations, that a sizeable percentage of the heat used in all heat-consuming mechanical devices is lost as a result of scattering or dissipation, and cannot be reused anew. Applied on a cosmic scale, Kelvin deduced that, in the long term, all energy will eventually be converted into dissipated heat — at which point not only all life but all movement in the universe will come to an end.

This first postulate was published in the *Proceedings of the Royal Society of Edinburgh* on June 19, 1852, and stated this ominous portent in the following words: "There is at present in the material world a universal tendency to the dissipation of mechanical energy."

The reservation hinted at in the words "at present" refers only to the fact that it is impossible for anyone to know for certain if such a "universal tendency to the dissipation of mechanical energy" might not be reversed or undone at some later date or in some other part of the universe. The title, however, does not express any such reservations about the inevitability of this process: "On a Universal Tendency in Nature to the Dissipation of Mechanical Energy."

Until the present no scientist has stepped forward to contradict the approach of this impending event. Nor has anyone demonstrated a willingness to suggest any sort of rewinding mechanism present in the universe that might delay or reverse its arrival. This imminent thermodynamic doomsday received a humorous nod in the subtitle of a book entitled *Entropy*, by the Dutchman J. Zernicke (1972). The subtitle reads *The Devil in the Passenger's Seat*. Whatever may occur in our lives or the universe as a whole, the devil of dissipation remains seated next to us in the passenger's seat, constantly reminding us of our inevitable dissolution in the warm-death that is his ultimate domain.

Warm-death refers to the final state of the universe when all potential sources of heat and every sun has cooled down, after which the universe as a whole will have become slightly warmer. If we were to take the ultimate end state of all of these sources, however, the term "cold death" would be more appropriate. It has also been used to describe this end state.

Warm-death. Cold death. These words imply that the natural

scientist has appropriated judgment day. It is in this way that we see the modern natural sciences in the grip of hubris, the overweening pride against which the ancient Greeks warned us. It is also shows them in the grip of *suberbia*, or pride, which the Middle Ages saw as the root of all evil.

Chapter IV

Events Synchronous with the Discovery of the Second Law

This chapter deals primarily with all the synchronous events that accompanied the discovery of the second main law. These events will help us to better understand the second law by pointing beyond its discovery toward a pervasive shift that encompassed the culture as a whole. For reasons that will become clear, I will not start my analysis with events surrounding Sadi Carnot's 1826 publication but will begin with cultural events surrounding the work completed in 1850 and 1852 by Clausius and Kelvin.

The Last of England

In 1852, the English painter, Ford Madox Brown (henceforth named just "Brown" as is customary in the relevant literature), settled down before his easel to paint the picture that graces the cover of this book. The painting forms a slight oval that is 82.5 centimeters high by 75 centimeters across. It was entitled *The Last of England.* At that time the title was readily understood by the general public to mean "The last immigrants from England." Brown made his own comments on the painting, in which he stated that it dealt specifically with the great British migration that reached its height in 1852.

By the standards of its time, Europe had grown overpopulated. Its numbers had swelled by 30 million in only 20 years. To give a rough idea of what this meant we might consider that between 1800 and 1850, the population of London grew from 800,000 to more than 8 million. Paris experienced a similar explosion, rising from 600,000 to 5 million inhabitants, and Berlin grew from less than 200,000 to almost 3.5 million. Emmigration was the result of this explosion, especially emmigration to the United States. The years 1851 through 1855 witnessed the emmigration of 350,000 persons per year from Britain alone. After 1855 the numbers decreased although they remained high for a long time after.

Brown considered emmigrating to India with his family, especially when a good friend of his left the country. In the end he decided against it. In the painting, however, he gives full reign to a depiction of his earlier intention and the ambivalent feelings surrounding his desire to leave his country. This is evident when we draw our attention more directly to the painting.

The man with the hat and overcoat is Brown himself. His wife sits next to him. The umbrella that Brown holds in his left hand cannot prevent the wind from blowing her shawl around his ears. That they are braving the cold cannot be doubted. Brown's face is ruddy and somewhat blotchy in color. He intentionally exposed himself to cold wind while painting this canvas so that he might better represent the effects of cold wind and slashing rain on human faces. The couple is holding each other's hands. Mrs. Brown's pale blue hand encloses the hand of her child held in the crook of her left arm.

There are plenty of further details to be noted on Brown's crowded canvas. We should pay particular attention to the top left quadrant of the painting which depicts the area behind Brown's back. It is not possible to form a completely coherent picture of what transpires there. The area is quite crowded and it is impossible to clearly distinguish everything even on a life-size reproduction. I myself have never seen the actual canvas but commentators who have stood before it at the Birmingham Museum of Art comment about the crowded muddle in this section of the painting. Luckily Brown left a description of what he intended to represent in that part of the painting.

A family should be visible. A father, mother, a daughter and their younger children of whom Brown writes, "make the best of things with tobacco pipe, apples, etc." Further back we witness the figure of a miscreant; "a reprobate," writes Brown, "who curses the fatherland with his fist to the shore, his old mother rebuking him while a jolly comrade, with a red face, encourages him with drunken words." All of this is depicted according to Brown himself, who added even more details.

Everyone's judgment is that there is too much depicted in such a cramped space. It is remarkable that this is the case, not only for this painting, but for many of Brown's other paintings as well. This complaint also carries over into the school of painting to which Brown belonged, the school that declared itself Pre-Raphaelite. The Pre-Raphaelites dominated English painting in the middle of the nineteenth century and have held an eminent position up to the present in English art as a whole. They continue to occupy this prominent position in the opinion of the general public, whatever nuanced judgments one may privately harbor about their romantic, prudish, sensuous and often kitschy canvases.

Brown did not belong to the inner circle of the Pre-Raphaelite school, but he wholeheartedly shared its principles. The Pre-Raphaelites adopted their name to proclaim their admiration for a manner of painting that was practiced in Italy prior to Raphael, who had lived from 1483 to 1520. Their ideals and vision found embodiment in the art of Giotto and Fra Angelico.

But let us examine the Pre-Raphaelite's constant effort to depict reality in all its details. Ironside, in his study entitled *Pre-Raphaelite Painters* (1948), wrote that they painted, "selecting nothing and rejecting nothing." When Holman Hunt, a central figure in the movement, was painting his canvas *Mayday* at Oxford, he climbed the Magdalena Tower every morning at five o'clock, in spite of his advanced age, in order to paint Oxford with every possible detail his eye could reach from that vantage point. In the same spirit Brown ushered himself and his wife into the cold wintry air so that her shawl and their frozen hands could be portrayed more vividly. During preparatory excursions for his painting he constantly sought out more details in order to bring more of them alive on his canvases.

Turner's "Pictures of Bits"

The Pre-Raphaelites were not the first to wear themselves out in the minute depiction of detail. Since 1880 the British painter William Turner had been advocating "pictures of bits." He produced paintings of tree trunks with every nuance of the bark depicted precisely, or of clouds rendered down to the very smallest detail the discerning eye could differentiate. In his book *Before Photography*, Galassi refers to these attempts as "bits of nature." The same author points toward a number of painters dedicated to the same rendition of detail in nature. As an example, he mentions Albrecht Dürer and his miniatures from the beginning of the sixteenth century. For instance, he cites Dürer's famous "great piece of turf" which is painted so precisely that a botanical expert would have no problem in identifying the individual species of grass included in it. The aim of Dürer's miniatures, however, was to create a definitive flora. He wished to depict plants as they appeared in the herbarium, which, not coincidentally, happened to originate in his time. Luca Ghini, who lived from 1490 to 1556, constructed the first herbarium.

Ghini was a professor of pharmaceutical Botany in Bologna, and later in Pisa.

The Pre-Raphaelites did not paint minutely precise pictures so that they could serve the science of botany. They simply wanted to invest their canvases with everything visible. Obviously this was an insurmountable and impossible task. It was too much.

Pre-Raphaelites and Photography

Because of their detailed excess, pre-Raphaelite canvases often remind one of photography. By cramming so many immigrants in a boat in *The Last of England*, Brown's canvas resembles a photographic reality that within its limited space cannot exclude any existing detail.

In trying to characterize Pre-Raphaelite painting in his book *On the European Art of Painting after 1850*, De Gruyter speaks of a realism that strikes modern viewers as almost photographic. Galassi, whom we mentioned earlier, refers to their com-position as "the syntax of photography."

The Photo

The photo gives us too much detail, too much image. So much so that the photographer often discovers things in the final photograph of which he had no notion at the time it was taken. This is noted by Newhall in his book *Historic Events* (1838–1939), can be attested by our own personal experiences. I once heard the story of a married couple on safari who had themselves photographed on the Serengeti plain even though they had been strictly prohibited from even momentarily leaving their Jeep. The picture was taken, and when it was developed later, the married couple and their photographer discovered that a couple of lions had been lying only a short distance away.

While the Impressionists resisted the overbearing inclusiveness of the photo, the Pre-Raphaelites were only all too pleased to portray crowded scenes. The Impressionists reached their full maturity a quarter of a century later than the Pre-Raphaelites. The latter, however, emerged alongside the first flowerings and early popularity of photography.

The Early Flowering and Popularity of Photography

The very first photo dates from 1824. The exposure time took eight hours. With such a long exposure time, it was impossible to take anyone's portrait, even if photographers had had recourse to the later devices that were invented to buttress and support aching arms, legs and feet. In fact, neither could any part of a city or landscape be expected to sit perfectly still for this period of time, and a city or landscape would doubtless reflect all the change in light produced by the moving sun. The first photo therefore depicted an interior with all of its myriad shadows.

Around 1850 an exposure time of one second was achieved. This is why photography began to flower after this date. The following salient years can be mentioned: In 1851, the first ever periodical devoted to photography was established in New York, and a similar journal was founded in Germany in 1854. The first association of photographers (Societé Heliographique) came into being in Paris in 1851. The first photographer's association in London started in 1853. All of these details can be found in Erich Stenger's 1950 book, *Siegeszug der Photographie in Kultur, Wissenschaft, Technik* (The victorious rise of photography in culture, science and technology).

Overpopulation

It is curious that this overcrowded medium would develop just as the great cities of western and middle Europe became overpopulated. This overpopulation gave rise to emmigration, which was mentioned in our discussion of Brown's overcrowded painting, *The Last of England*.

In this book we have a caricature with the title *Overpopulation*, made in 1851 by George Cruikshank, who was a famous caricaturist of that time. This drawing presented a future London. A caricature, for sure, but a caricature which nevertheless conveyed a fear of an enveloping multitude, which had never been envisaged before. Independently of Cruikshank, Gustav Doré also produced a picture of London with no holds barred. One sees a nightmarish mass of people. A caricature of a nightmare. But in the meantime, all of the great cities of Europe were taking on this nightmarish quality. This overcrowding was not only evident in very large cities; it eventually extended to smaller cities and the traffic between them. When we currently think of overcrowding, we are likely to think of crowded freeways.

In my youth my parents could travel safely by bike on the main road connecting my native town of Deventer and Apeldoorn. There was so little traffic that they could ride their bikes in the middle of the road until one of them yelled "car coming!" After the lone car passed them and vanished in its accompanying cloud of dust, my parents owned the road once more.

Let us examine Cruikshank's etching again. Even the roofs of its teeming shelters are occupied. Captions clearly indicate that living space is sought. On the lower right there is a party sitting at a table drinking beer. This is a familiar sight to moderns in our time because with current overpopulation we are accustomed to seeing drinking and eating in public. Even in the street, we witness people drinking from a can, or eating out of a bag.

Another odd shift can be witnessed on Cruikshank's print. Not a single priest or clergyman is in evidence. There are no noblemen, either. The first and second estates are entirely missing from the picture. Their time had passed.

George Cruikshank's *Overpopulation* (1851).

The New York Statue of Liberty

When the United States was celebrating the first centenary in 1876, France gave the youthful celebrant the Statue of Liberty, which was placed on a little island before the city of New York. Many Europeans welcomed the sight of this statue as the promise of freedom. One only needs to read *My Life* by Golda Meir to realize how much this monument represented the promise of freedom to immigrants, and how much of that promise was eventually fulfilled for many of them. The persecuted were not the only ones to flee to the New World. Many were simply oppressed, exploited, and discouraged by the socioeconomic circumstances fostered by the rapid industrialization of Europe. They were the pauperized surplus population that Europe no longer had any place for.

All of this can be gleaned from the monument's inscription, which was written by Emma Lazarus and engraved on the statue in 1903. It reads: "Keep, ancient lands, your storied pomp! Give me your tired, your poor, your huddled masses yearning to breathe free, the wretched refuse of your teeming shore."

Yet, in 1903, the United States had already been trying to stem the tide of immigrants for nearly ten years, as the historian and scholar Petersen notes in his book *The Technical Labyrinth* (1981). Among these masses of millions, the number of those belonging to the first and second estates was negligible. Those belonging to the third estate had grown out of all proportion, and they formed the "masses" that crossed the great waters from "Europe's teeming shores." Released from the ordering influences and constraints of the first and second estates, the third estate grew into a teeming mass.

The Machine

This study has already noted that the French Revolution was responsible for declaring the third estate, the estate of the horde, to be the only estate worth preserving. This judgment was subsequently validated for all of Europe and the world as a whole. All reactions have failed to undo this declaration and its consequences, and will never succeed in doing so. This is because the third estate finds its origins in the machine, which was invented in the eighteenth century. These machines were powered not only by the heat produced by the newly commodified wood, but also by the newly created masses of people released from the bonds of clergy and nobility. It was the machine that gave birth to the factory and to the laboring masses of the third estate. None of these will disappear until all mineral and fossil fuel is exhausted. If these fuels are eventually replaced by another artificial "fuel," the regime of the machine will continue on its course until the earth risks becoming uninhabitable. Up to now, however, such a new energy source has been merely a wishful thought. If that wish does not become a reality and if a new and limitless source of energy is not found, it is conceivable that mankind might be forced to return to an older agrarian model of self-sustaining societies. If this were to come about, the two laws of thermodynamics would no longer play such a central role in our life. Strange as this may sound, these laws would then have become obsolete.

Marginal, Dissipated and Scattered Humanity

Both laws currently prevail as if they were eternal. The first law remains valid because we are still under the sway of metaphysically transformed substances. We continue to consume universalized bread and make use of universalized elements. The second law remains valid because mankind is presently a universal and teeming multitude, a mass. In the mid-nineteenth century this teeming mass spilled over from the shores of Europe in the manner suggestive of the scattering that attends the process of converting heat into mechanical energy. These newly created masses become scattered and dispersed over the entire globe.

It is this phenomenon of popular dispersal that we observed in the Brown painting, *The Last of England*, depicting a boat of emmigrants so crowded with people that they appear to risk drowning. Such dispersal and crowding can now be seen on almost any photograph because this manner of representation by its very nature shows us *too much*.

Sadi Carnot

It will be useful for us to return once more to Sadi Carnot's book, *Reflections on the Driving Force Inherent in Fire and on the Machines Capable of Using This Power.* The book prefigures the discovery of the second law by stating that heat converts to power through a process that proceeds exclusively in a single direction from hot to cold. Carnot mentions nothing about dissipation in this book. It was only in papers that came to light after his death that note is made of this particular phenomenon. By this time, however, the articles by Clausius and Kelvin had already been published.

Carnot's book was written in 1824. This once again raises the question: What events were synchronous with the publication of Carnot's book? Naturally, a great deal happened in 1824. What is most surprising, however, is that this is the year in which *the first photograph* was taken. 1824 was the year of the first photo. Many disputes have arisen ever since concerning the exact date of the first photograph. This is because many individuals were working independently of one another on the discovery of photography around this time. This gave rise to a dispute over who could be rightfully declared the first inventor of this process. According to an article by E. Chevren, the true inventor was Nicéphore Niépce and his first photo was dated 1824.

In his great work, *Geschichte der Photographie*, better known as the *History of Photography* in its Dover edition, J. M. Eder also named Niépce as the inventor but wishes to date the first photo exactly two years later — for reasons that do not seem entirely solid to me. But even if the date 1826 turns out to correct, the closeness of Niépce's discovery and Carnot's publication is striking.

The exact nature of the synchronicity between the discovery of photography and the second main law has already been explained in relation to the authors Clausius and Kelvin. It is unnecessary to repeat it again fully, but we can summarize it in the following sentences. Photography gives us too much; it provides an excess of detail above and beyond what is required in the image. In the second law of thermodynamics, this same excess is directly expressed through the phenomenon of dissipation. A surplus of heat is produced in the process of powering a machine. A surplus of detail is produced in the photographic transference from light to image.

An Aside

If has often been noted that photography could have been invented considerably earlier. The "dark room" or "camera obscura" was already generally known in the sixteenth century, and was even used in portable form by draftsmen and painters in the seventeenth century. In 1727 Heinrich Shultze discovered the chemical material needed to fix the image from a camera. Looking at these pieces of evidence, Helmut and Alison Gersheim, who are known for their great photohistorical collection in Austin, Texas, have exclaimed, "that photography was not invented earlier remains the greatest mystery of its history." A mystery, we may add, that loses much of its mysteriousness once one takes Carnot's book into consideration, and does not hesitate to draw a connecting line between them.

Second Aside

Having mentioned the Gersheims and their beautiful published collections of old photo portraiture, I do not want to omit my admiration for them. From what I have said in previous paragraphs, the reader may get the impression that I reject all photography. This is not so. I find that they can be splendid, provided they do not flaunt their excessive detail, their "too much." This holds for portrait photos especially. Another question has been raised many times as to why early daguerreotypes and portrait photos present subjects with such imposing and hauntingly honest faces. In *Life in Plurality and Complex Society* (1974), I attempted to answer that question.

Chapter V

Transubstantiation

We saw how the first law of thermodynamics was discovered in the wake of the French Revolution. We described how Count Rumford came to an understanding of this law while supervising the production of cannons at a munitions factory in Munich.

A full account of this historical discovery must include the fact that it took place in a world that had been radically altered by the events surrounding the French Revolution. We noted that Rumford lived at a time when the cultural as well as the material world was changed in ways that came only gradually to light in subsequent cultural and scientific developments. We take as our point of departure the assumption that radical historical shifts do not merely change human minds but that they also change the matter contemplated by these minds. Rumford lived at a time of significant historical change when metal underwent a *transubstantiation*. He faced a new metal that demanded to be understood in new ways.

Some will no doubt object to the use of the word "transubstantiation" within this context. Historically the term applies to a specifically Christian sacred ritual in which bread and wine is miraculously changed into the body of Christ. To compare the change that took place in Rumford's metal to the central mystery of the Church might appear inappropriate or even sacrilegious. We therefore should perhaps distinguish a sacred transubstantiation, which implies an encounter between heaven and earth, from a profane transubstantiation that reveals itself in the course of ordinary history.

Let us briefly trace the history of the concept of sacred transubstantiation. It refers originally to a miraculous change that takes place within the context of a religious remembrance of Christ's death and resurrection. It came into use in the twelfth century and it bears in its etymology the traces of an Aristotelian understanding of a thing as made up of an underlying core or foundation (L *sub-stancia* or G. *hypo-keimenon*) supporting the various properties implanted upon it.[1]

It is interesting to note that in the course of the last century, in particular, the term *transubstantiation* has come under increased theological scrutiny and criticism because it tended to emphasize the aspect of a miraculous change in the host and sometimes led to a comparative neglect of the commemorative aspect of the sacrament. During the Last Supper Christ had specifically instructed his disciples to institute a commemorative meal and to eat bread and drink wine in memory of him. The Last Supper became the mystical gathering of the faithful in thankful memory of the meal that took place in the cenacle, just outside of Jerusalem.

---

[1] Martin Heidegger, *What Is a Thing* (Chicago: Henry Regnery, 1969), 34.

The *Dictionnaire de la Spiritualité*

This encyclopedia of spirituality devotes 16 hefty tomes to an exhaustive treatment of virtually everything of importance concerning Catholic doctrine in the twentieth century. Yet, quite significantly, there is no separate entry in this encyclopedia under the heading of "transubstantiation." All matters pertaining to this topic are treated under the heading of "Eucharistie," the Eucharist. The encyclopedia devotes more than 100 columns to this important topic.

The word *Eucharist* derives from the Greek *Eucharistos* for "grateful" or "thankful." The verb *euchariteoo* is translated as "to be thankful," "to return thanks" and "to requite." This choice of words to describe the high point of the mass derives its spiritual authority from Christ's own words, spoken at the Last Supper: "Do this in memory of me." The word as such draws our attention to the thankful and commemorative aspect of the ritual and places a lesser emphasis on the aspect of a miraculous transubstantiation. In time this de-emphasis tended to take the form of neglect and it became necessary to warn modern believers not to fall into the error of understanding the Eucharist purely as a commemorative ritual and to overlook the aspect of the real presence (*preasentia realis*) of the body of Christ. This doctrinal correction makes it clear that even though we may avoid using the word *transubstantiation* we should not overlook the aspect of a miraculous transformation that remains an intrinsic part of the ritual of the Eucharist.

The Reformation

The Reformation took its distance from the traditional idea of a real presence or *preasentia realis* in the Eucharist. Luther at first hesitated, but then came to reject the traditional idea and the other reformers followed suit. The reformers were not unmindful of the biblical phrase: "This *is* my body, this *is* my blood." They rather sought to replace the word "is" with some other word less suggestive of a miraculous transubstantiation, such as: "This *means* my body and my blood," or "This *refers to* my body and my blood."

It goes without saying that the Reformation theologians developed elaborate doctrines to support their new reading of the biblical texts. Their doctrines embodied earnest thought and serious intentions and they can still be read today with profit. This is not the place to explore in depth a complex theological controversy. Our sole purpose here is to point out that during the first half of the sixteenth century the traditional doctrine of transubstantiation with its implication of a miraculous and substantial change from bread and wine into the body of Christ came to be questioned by a growing number of Christian believers.

To better understand this change in belief and perception in the realm of Christian theology we must turn to similar changes and developments in other cultural realms that occurred at about the same time. As we recall, our exploration of the two laws of thermodynamics followed a similar path and examined discoveries and innovations in the fields of physics and chemistry in the light of developments that occurred nearly simultaneously in the realms of politics and economics. We proceed here in a similar fashion by comparing significant innovations in the realms of art, medicine and astronomy that occurred at the same time as when the Christian world began to develop significant alternatives to the traditional doctrine of transubstantiation.

Perspective

Sebastio Serlio published in 1537 his *Libri dell'architecttura* in which he presents a pictorial representation of a city that in all respects exemplifies a strict and mathematical application

Sebastio Serlio, *Libri dell'architecttura* (1537).

Albrecht Dürer, *Unterweisung der Messung* (1538).

of the principles of central perspective. All the vertical lines of the drawing converge upon a black dot at the very center of the plate. This single black dot represents the very heart and soul of the pictorial presentation and the place from which everything begins and ends. The author carefully explains how he has gone about constructing every line by relating it to this dot.

We are somewhat surprised at the sterile quality of the drawing and its strange lifeless appearance. Serlio's city appears to reject human encounters and leaves no room for intimacy. It appears unable to make place for an individual, inhabited house with its own inherent logic and its own way of obeying and belonging to a larger world. We see instead a town where every nook and cranny has been deprived of its individuality and has been ruthlessly incorporated into the abstract mathematical order of the central perspective. A space designed on such principles remains effectively uninhabitable. It represents a gloomy, self-sufficient, singular world that no longer seeks contact with any other; it leaves no room for intimacy or for the miracle of a true encounter. It also leaves no place for a sacred transubstantiation in the manner in which it had been traditionally conceived. We may understand the Reformation as an attempt to conform religious belief to the new realities of a space of time that was being transformed by the logic of a central perspective. Serlio's woodcut represents an imaginary city that was in the process of being built. It announces a world that is in the process of being changed in such a way that it no longer leaves any room for miracles.

It is interesting to compare Serlio's treatise with that of Albrecht Dürer's *Unterweisung der Messung* (About measurement), which appeared in 1538. This book contains equally precise instructions on how to depict three-dimensional space with the aid of central perspective. It provides a number of illustrations, one of which we reproduce here. The artist is seated at a table from where he observes his model reclining on a bed. He keeps his eye glued to a tiny obelisk placed before him on the table. This small object serves the purpose of keeping the eye of the artist fixed into one position and thus keeps it from wandering outside the frame imposed by the central perspective.

A transparent screen or veil, a *velo*, is placed between the fixed eye of the artist and the reclining model in front of him. The *velo* is geometrically divided by equally spaced horizontal and vertical lines and these lines are exactly reproduced on the sheet of paper on which the artist draws. The artist observes what is visible on the screen strictly from the position of the obelisk. He then draws what he sees at the exactly corresponding location on the paper before him. In this way he is able to reproduce with mathematical exactitude a three-dimensional figure on a two-dimensional sheet of paper.

Everything begins and ends here at the same fixed point. There is no room in it for an exchange of glances, for a true meeting and a mutual revelation of self and other. The pictorial space cannot make room for spontaneous gestures, surprising events or miraculous manifestations.

We should also pay close attention to the reclining woman in Dürer's print. She lies on a bed supported by two cushions but she finds herself in anything but a comfortable position. With her right hand she grips a sheet to cover part of her body and her sex, yet her breasts remain fully exposed. She cannot hide from the fixed eye of the observer and her feeble attempts at modesty appear quite ineffective. The pictorial space deprives her all at once of her ability to hide and to freely reveal herself. One gains the impression that her seminudity leaves her more fully exposed to the clinical eye of the artist and the viewer than would complete and frank nudity.

We cannot help being reminded of Serlio's cityscape, which was constructed on similar mathematical principles. Both Dürer's model and Serlio's city appear captured rather than fully revealed within the strangely lifeless geometric time and space of the central perspective. With these pictorial representations we seem to have entered a new world in which there is no longer room for intimate encounters or for miraculous transformations.

Copernicus's Revolution

Still in the same sixteenth century, in 1543, Copernicus published his *De Revolutionibus orbium coelestium* (About the rotation of heavenly bodies). The author had long hesitated to publish his revolutionary insights and only on his deathbed was the author able to examine the proofs. His publisher had taken the precaution of presenting the author's views as a hypothesis, rather than as a verifiable scientific reality.

The new cosmology, which replaced the earth as the very center of a planetary system, represented a serious threat to the established world order. By contrast, the pictorial technique of artistic perspectography appears to have aroused little or no controversy. Yet the two discoveries appear closely interrelated and cannot be fully understood without reference to each other.

Serlio and Dürer introduced a technique of spatial and temporal representation that in its extreme and mechanical application drains a reclining nude of all its mystery and transforms a living city into a virtual ghost town. Considered by itself and absent of any countervailing force Copernicus's astronomical vision had a similar power to empty the heavens of its mystery and render it as lifeless as Serlio's uninhabitable city or Dürer's objectified nude.

It appears that at the time of Serlio, Dürer and Copernicus a radical change took place in the Western world that took the form of a profane transubstantiation. This profane metamorphosis had the effect of altering the very substance of every earthly and heavenly thing. It would create a perspective that could ban intimacy and revealing encounters from the world. It could make a city look uninhabited and even uninhabitable. It could cast doubt and suspicion on an ancient article of faith and ban miracles from the world.

Conclusion

Our exploration of the theme of transubstantiation has shown that every significant innovation or discovery can be approached from the side of the artist or scientist who proposed and elaborated it. An important discovery brings about a change in the outlook and understanding of the one who makes the discovery. It can also be approached from the side of the newly elaborated or discovered matter as announcing a change in the established order of the natural world. It is possible for us to imagine the process of natural scientific discovery as a progressive revelation of the nature of material relations. It is also possible to assume that the natural world, like the cultural world, undergoes periodic changes that are noticed and recorded by observant scientists. In this way we come to think of every important scientific discovery as confronting us with matter metamorphosed or transubstantiated in either a profane or a sacred sense.

Afterword

Thomas Kuhn, the author of *The Structure of Scientific Revolutions*, published an article in 1959 on the discovery of the law of the conservation of energy. In this article he makes no mention of Rumford but he does cite Robert Mayer, who later discovered the same law independently of Rumford and published his findings in 1842. Kuhn admits that Mayer was not the only discoverer of the law since James Joule (1843), Colding (1851) and Helmholtz (1847), along with a few others, can all lay legitimate claim to the same discovery as well. In total, 12 men are mentioned as having discovered the law independently of one another. Kuhn is justifiably astonished. He asks how it could be that within a relatively short time frame so many experimental findings and conceptual developments converged and began to point in the direction of a completely new understanding of matter and energy.

Kuhn was unable to find a satisfactory answer to this question. To begin to understand the problem he posed we need to look beyond the circumscribed field of natural science and examine the broader context of human events as these unfolded at the time of the revolutionary developments in physics.

No science is autonomous. Every scientific investigation takes place within a larger world of human activities, of prevailing attitudes, thoughts and feelings. Some have suggested that science and technology should be understood as autonomous forces that by themselves originate new modes of thought and create sweeping changes in the ways we understand ourselves and our world.

Nonetheless, this is not the viewpoint defended in this essay. We have suggested that in Rumford's time metal and matter underwent a fundamental change. But we also mentioned a certain Count Rumford who observed the change, interpreted it and made it part of the human world. The same counts for Mayer's discovery. We postulated that in Mayer's time, seawater became subject to an important change. But that change needed to be observed, thought through and recorded.

We do not think of scientific discovery as simply the product of material changes in a material world. Nor do we attribute it simply to genius or to the powers of the human mind alone. We presented scientific discovery as a *revealing coming together* of an incarnated human presence and a changing natural world. Technology contains within itself a force and momentum that in part determines the path it follows in the human world. It has a force and an authority all its own. But there can be no science and no discovery of natural things or forces without a revealing encounter between the two. There can be no question of a scientific discovery without such a conjoining of person and thing.

Bibliography

Atkins, P. W. *The Second Law*. New York, 1984.

Bachelard, G. *La psychanalyse du feu*. Paris, 1946.

Binkley, R. C. *Realism and Nationalism, 1852–1871*. New York, 1963.

Briggs, A. *The Age of Improvement*. London, 1980.

Brown, S. C. *Count Rumford, Physicist Extraordinary*. New York, 1926.

Bury, J. P. T. *The Zenith of European Power*. Cambridge, 1964.

Cardwell, D. S. L. *From Watt to Clausius: The Rise of Thermodynamics in the Early Industrial Age*. London, 1971.

———. *Technology, Science and History*. London, 1972.

Carnot, S. *Réflexions sur la puissance motrice du feu et sur les machines propres à développer cette puissance*. 1824; reprint, Paris, 1912.

Chevreul, E. "La vérité sur l'invention de la photographie." *Journal des savants* (February 1873).

Cicero. *The Nature of Gods (De natura deorum)*. London, 1972.

Clausius, R. "Ueber die bewegende Kraft der Wtirme and die Gesetze, welche sich daraus für die Wärmelehre selbst ableiten lassen. *Annalen der Physik und Chemie* 79 (1850).

———. *Ueber den zweiten Hauptsatz der mechanischen Wärmetheorie*. Braunschweig, 1867.

Clough, S. B., and C. W. Cole. *Economic History of Europe.* Boston, 1952.

D'Asao, A. *The Rise of the New Physics.* 2 vols. Dover, 1951.

Eder, J. M. *History of Photography.* New York, 1978.

Elkana, Y. "The Conservation of Energy: A Case of Simultaneous Discovery?" *Archives internationales d'histoire des sciences* 23 (1970): 90–91.

Ellis, G. E. *Memoir of Sir Benjamin Thompson, Count Rumford, with Notices of His Daughter.* Philadelphia, 1871.

Folie, F. "R. Clausius: Sa vie, ses travaux et leur portée métaphysique," *Revue des questions scientifiques* 27 (1890).

Galassi, P. *Before Photography: Painting and the Invention of Photography.* The Museum of Modern Art, 1981.

Gernsheim, H. *The Origins of Photography.* London, 1982.

Gernsheim, H., and A. Gernsheim. *Historic Events, 1839–1939.* New York, 1949.

———. *L. J. M. Daguerre.* London, 1956.

Giedion, S. *Space, Time and Architecture.* 1941; reprint, Cambridge, Mass., 1966.

Gruyter, W. J. de. *De Europese schilderkunst na 1850.* Den Haag-Antwerpen, 1954.

Heidegger, M. *Die Technik and die Kehre.* Pfullingen, 1962.

Hell, B. *Robert Mayer und das Gesetz von der Erhaltung der Energie.* Stuttgart, 1925.

Hilton, T. *The Pre-Raphaelites.* London, 1970.

Hobshawn, E. J. *The Age of Capital.* London, 1976.

Hooykaas, R. *Geschiedenis der Natuurwetenschappen.* Utrecht, 1971.

Ironside, R. *Pre-Raphaelite Painters.* London, 1948.

Jay P. *Nicéphore Niépce.* Paris, 1983.

Klemm, F., and H. Schimank. "Julius Robert Mayer zum 150 Geburtstag." *Deutsches Museum, Abhandlungen und Berichte* 33 (1965): 3.

Kuhn, Th. S. "Energy Conservation as an Example of Simultaneous Discovery." In *Critical Problems in the History of Science.* Madison, Wis., 1959.

Laver, J. *The Age of Optimism, Manners and Morals, 1848–1914.* London, 1966.

Mach, E. *Die Principien der Warmelehre historisch-kritisch entwickelt.* Leipzig, 1900.

Marx, K. "Zur Kritik der politischen Oeconomie." In Karl Marx and Friedrich Engels, *Werke*, vol. 13. 1859; reprint, Berlin, 1978.

Mayer, J. R. "Bemerkungen über die Krâfte der unbelebten Natur." *Annalen der Chemie and Pharmacie* 42 (1842).

Merz, J. Th. *A History of European Scientific Thought.* Vol. 2. New York, 1965.

Mittasch, A. Julius *Robert Mayers Kausalbegriff.* Berlin, 1940.

Pevsner, N. *Art and Architecture in the Zenith of European Power.* Cambridge, 1964.

Picard, E. *Sadi Carnot: Biographie et manuscrit.* Paris, 1927.

Pieterson, M., ed. *Het technisch labyrint.* Leiden, 1981.

Proudhon, P.-J. *Qu'est-ce que la propriété?* Paris, 1840.

Rifkin, J. *Entropy: A New World View.* New York, 1980.

Roller, D. *Count Rumford (Benjamin Thompson).* Discourse before the Royal Society, 1798.

Romanyshyn, R. D. *Technology as Symptom and Dream.* New York, 1989.

Rumford, Benjamin, Count of. *An Inquiry Concerning the Source of Heat Which Is Excited by Friction.* Read before the Royal Society, January 25, 1798.

Schmolz, H., and H. Weckbach. *Robert Mayer, sein Leben and Werk in Dokumenten.* Weiszenhorn, 1964.

Smith, C., and M. Norton Wise. *Energy and Empire: A Biographical Study of Lord Kelvin.* Cambridge, 1989.

Stenger, E. *Siegeszug der Photographie.* Seebruck am Chiemsee, 1950.

Tait, P. G. *Sketch of Thermodynamics.* 1868; reprint, Edinburgh 1877.

Thomson, D. *Europe since Napoleon.* London, 1957.

Thomson, W. (Lord Kelvin). "On a Universal Tendency in Nature to the Dissipation of Mechanical Energy." Paper read before the Royal Society of Edinburgh, April 19, 1852.

Wallace, A. R. *The Wonderful Century.* 1898; reprint, London, 1970.

Watanabe, M. "Count Rumford's First Exposition of the Dynamic Aspect of Heat." *Isis* (1913).

Weyrauch, J. J. *Robert Mayer, Kleinere Schriften and Briefe.* Stuttgart, 1893.

Wilson, M. "Count Rumford." *Scientific American* 203 (1960).

Wolf, A. *A History of Science, Technology and Philosophy in the Eighteenth Century.* Vol. 1. New York, 1961.

Zernike, J. *Entropy, the Devil on the Pillion.* Deventer, 1972.

Index